ELEMENTS

OF

ANALYTICAL GEOMETRY.

BY

ALBERT E. CHURCH, A. M.,

PROFESSOR OF MATHEMATICS IN THE U. S. MILITARY ACADEMY; AUTHOR OF
ELEMENTS OF THE DIFFERENTIAL AND INTEGRAL CALCULUS.

SECOND EDITION

NEW YORK:
PUBLISHED BY A. S. BARNES & BURR,
51 & 53 JOHN STREET.
SOLD BY BOOKSELLERS, GENERALLY, THROUGHOUT THE UNITED STATES.
1862.

PREFACE.

No branch of pure Mathematics presents more to interest and improve the mind of the mathematical student, than Analytical Geometry. Uniting the clearness of the geometrical reasoning, with the brevity and generality of the algebraic, it not only satisfies the requirements of the closest reasoner, but gives continued and increasing pleasure, by the elegance with which its varied results are deduced and interpreted.

In preparing this treatise the Author has endeavoured to preserve the true spirit of Analysis, as developed by the celebrated French mathematician, Biot, in his admirable work on the same subject, while he has made such changes, both in the arrangement of the matter and the methods of demonstration, as he believed would render the whole more attractive, and easily acquired by any student possessing a knowledge of the elementary principles of Algebra and Geometry.

In discussing the Conic sections he has preferred to consider the Parabola first, not only for the reason that the properties of this curve are more simple and more easily deduced than those of the others, but because, by this course, he was enabled to treat of the Ellipse and Hyperbola together, thus avoiding much of the repetition of words, which necessarily arises from their separate discussion.

Although the treatise has been prepared with special reference to the wants of the Author's own classes at the Military Academy, he trusts that it will be found acceptable and useful to all, who are disposed to advance in the higher branches of Analysis.

Those who desire to make the subject as practical. as may be, will find in the last article of the work a large number of examples.

U. S. Military Academy,
West Point, N. Y., July 1, 1851.

CONTENTS.

PART I.

DETERMINATE GEOMETRY.

PART II.

INDETERMINATE GEOMETRY.

ANALYTICAL GEOMETRY.

PART I.

DETERMINATE GEOMETRY.

1. GEOMETRY, in its most general sense, has for its object, not only the measurement, but the development of the properties and relations, of lines, surfaces, and solids.

This object may be attained, either by operating directly upon the magnitudes themselves; or, by representing them and their parts, by algebraic symbols, and operating upon these representatives by the known methods of Algebra, thus deducing results essentially the same as those which would be obtained by the direct method. As the reasoning employed is much generalized, and operations are much abridged by the application of Algebra, the latter method evidently possesses many advantages over the former.

This latter method, which is *Analytical Geometry*, may be defined to be: *That branch of Mathematics, in which, the magnitudes considered are represented by letters, and the properties and relations of these magnitudes made known by the application of the various rules of Algebra.*

Analytical Geometry may be *Determinate*, or *Indeterminate*.

2

Determinate, when it has for its object the solution of determinate problems, that is, of problems, in which, the given conditions limit the number, and afford the means of deducing the values, of the required parts.

Indeterminate, when it has for its object the discussion of the general properties of geometrical magnitudes.

2. Geometrical magnitudes may be represented algebraically, in two ways.

First. The magnitudes may be directly represented by letters; as the line AB, given absolutely, may be represented by the symbol a. Likewise, the square AC, may be represented by the symbol A; or better by the symbol a^2, a being the representative of the side AB. Also, the rectangle ABC′D′ may be represented by the symbol B; or by the product ab, a and b being the representatives of the sides AB ana BC′; or more properly by c^2, c representing the side of a square equivalent to the rectangle. In the same way, a cube would be represented by a^3, a being the representative of one of the edges; and a rectangular parallelopipedon by abc or by d^3.

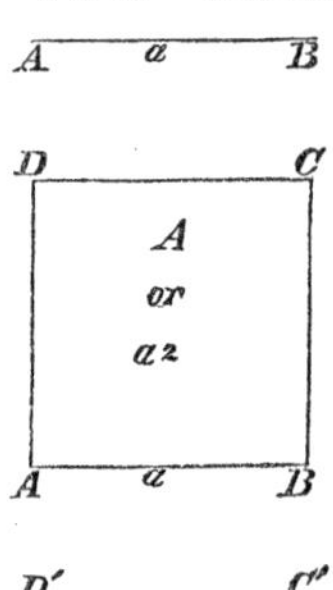

And in general, we thus represent a definite portion of a line, whether straight or curved, by a single letter or expression of the first degree; a surface by the product of two letters or an expression of the second degree; and a solid by an expression of the third degree.

Second. Instead of representing the magnitude directly, the algebraic symbol may represent the number of times, that a given or assumed unit of measure is contained in the magnitude; as, for the line AB, a may represent the number of times that a known unit of length is contained in it; and a^2 and ab or c^2, the number of times that a square whose side is the unit of length, is contained in the given square or rectangle; and a^3 and abc, the number of cubic units contained in the given cube or parallelopipedon.

Since, in this case, the algebraic symbols represent abstract numbers, any algebraic expression, thus composed, is called *an abstract expression* or *equation*, to distinguish it from one in which the direct representatives of the magnitudes enter. Since a line is always represented by *an algebraic expression of the first degree*, such expression is called *linear*. Also, *a linear equation is an equation of the first degree.*

3. From what precedes, it is evident, that any abstract expression may be changed into one in which the direct representatives of the magnitudes enter, *by substituting, for the representative of each abstract number, the representative of the magnitude divided by the representative of the unit of measure.* Thus in the expression,

$$x = a + b,$$

x, a and b representing numbers; if we substitute for them, their equals $\frac{X}{l}$, $\frac{A}{l}$, $\frac{B}{l}$, X, A and B being the direct representatives of the magnitudes, and l that of the unit of measure, we have

$$\frac{X}{l} = \frac{A}{l} + \frac{B}{l} \qquad \text{or} \qquad X = A + B.$$

In the same way, the abstract expression

$$x = ab + c,$$

may be changed into the corresponding one,

$$\frac{X}{l} = \frac{A}{l}\frac{B}{l} + \frac{C}{l} \qquad \text{or} \qquad Xl = AB + Cl.$$

It should be remarked, that every expression of this kind must be homogeneous, else we should have magnitudes of different kinds added or subtracted or equal, which can not be.

4. After having deduced a result, by the application of algebra to a geometrical proposition, it will be necessary to explain this result geometrically, that is, *to draw a geometrical figure, each of the parts of which shall have its representative in the algebraic expression, and also have the same geometrical relation to the others, as that indicated in the expression.* This is called *constructing the expression.*

Examples.

1. Let $$x = a + b.$$

If a and b are the direct representatives of right lines, x will be the representative of their sum. To construct it, take the line represented by a, in the dividers, and from any point A, on the indefinite line X′X as a point of beginning, or origin, lay off AB equal to this distance, then from B lay off BC equal to the line represented by b, the line AC $=$ AB $+$ BC will evidently be represented by x.

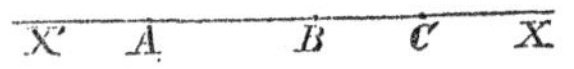

X′ A B C X

Or if a and b represent numbers, lay off from A, a times the unit of length, then from B, b times the same unit, and, as before, AC will be the line represented by x.

2. Let $$x = a - b.$$

From A lay off AB $= a$, then from B lay off, towards A, the distance BC $= b$;

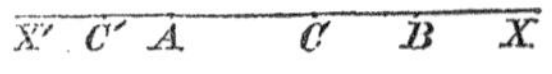

X′ C′ A C B X

$$\text{AC} = \text{AB} - \text{BC}$$

will be the line represented by x.

If $a = b$, x will be equal to 0, the point C will evidently fall on A, and there will be no line.

If $b > a$, x will be essentially negative, the point C will fall on the left of A, as at C′, and AC′, laid off from A to the left, will be represented by x. Thus, we see an illustration of the principle taught in Trigonometry, that if lines having the positive sign are estimated or laid off in one direction, those having the negative sign must be estimated in a contrary direction.

3. Let $$x = \frac{ab}{c}.$$

In this case x is a fourth proportional to c, a and b, and is thus constructed. Draw any two right lines making an angle; on one, from their point of intersection, as an origin, lay off the distances AC $= c$ and AB $= a$; on the second, lay off AD $= b$; join the points C and D, and through B draw BX parallel to CD; AX will be the line represented by x. For, we have

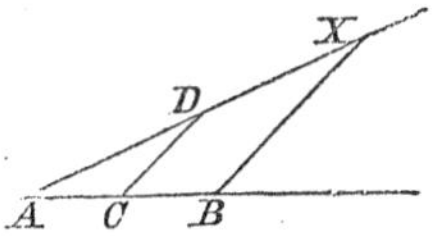

$$\text{AC} : \text{AB} :: \text{AD} : \text{AX} \quad \text{or} \quad c : a :: b : \text{AX}$$

whence

$$\text{AX} = \frac{ab}{c} = x.$$

4. Let $$x = \frac{abc}{de}.$$

This may be put under the form

$$x = \frac{ab}{d} \times \frac{c}{e}.$$

Place $\frac{ab}{d} = g$, and construct g as above, then we have

$$x = \frac{gc}{e},$$

which may be constructed in the same way; and so with any expression, in which the number of factors in the numerator is one greater than in the denominator.

5. Let $x = \sqrt{ab}$ or $x^2 = ab$.

In this case, x is a mean proportional between a and b. To construct it: On any right line, lay off AB $= a$; from B lay off BC $= b$; on the sum, AC, describe a semi-circle, and at the point B erect BX perpendicular to AC. The part BX, included between the diameter and circumference, will be the line represented by x. For from a known property of the circle, we have

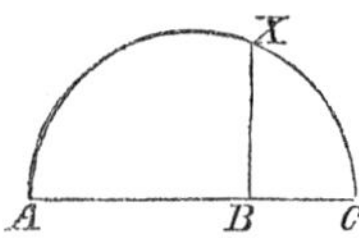

$$\overline{BX}^2 = AB \times BC \quad \text{or} \quad BX = \sqrt{ab} = x.$$

6. Let $x = \sqrt{\frac{abc}{d}} = \sqrt{\frac{ab}{d} \times c}.$

Place $\frac{ab}{d} = g$ and construct it as in example 3, then we have

$$x = \sqrt{gc},$$

which may be constructed as above.

7. Let $x = \sqrt{a^2 + b^2}$ or $x^2 = a^2 + b^2$.

In this case, x is the hypothenuse of a right angled triangle, the two sides of which are a and b. Therefore, draw two lines forming, with each other, a right angle: From the vertex, A, on one, lay off AB $= a$; on the other, lay off AC $= b$; join B and C, the line BC will be represented by x. For we have

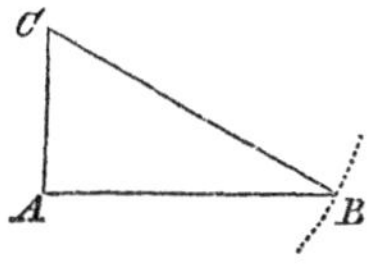

$$\overline{BC}^2 = \overline{AB}^2 + \overline{AC}^2 \quad \text{or} \quad BC = \sqrt{a^2 + b^2} = x.$$

8. Let $$x = \sqrt{a^2 - b^2}.$$

From A, in the last figure, lay off AC $= b$; then from C as a centre, and with CB $= a$ as a radius, describe an arc cutting AB in B; the distance, AB, will be represented by x. For

$$AB = \sqrt{\overline{BC}^2 - \overline{AC}^2} = \sqrt{a^2 - b^2} = x.$$

9. Let $$x = \sqrt{a^2 + b^2 - c^2}.$$

Place $a^2 + b^2 = g^2$, and construct g as in example 7; then we have

$$x = \sqrt{g^2 - c^2},$$

which may be constructed as above.

10. Let $$x = \sqrt{a^2 + ac}.$$

11. Let $$x = \frac{abc + g^2d}{fc}.$$

12. Let $$x = \sqrt{\frac{a^2c}{b}} + \sqrt{a^2 - bc}.$$

5. Let us now construct the roots of the four forms of equations of the second degree. *The first,*

$$x^2 + 2ax = b^2,$$

gives the roots

$$x = -a + \sqrt{b^2 + a^2} \qquad x = -a - \sqrt{b^2 + a^2}.$$

From any point, as A, lay off AB $= b$; at B, erect the perpendicular BC $= a$, then as in example 7

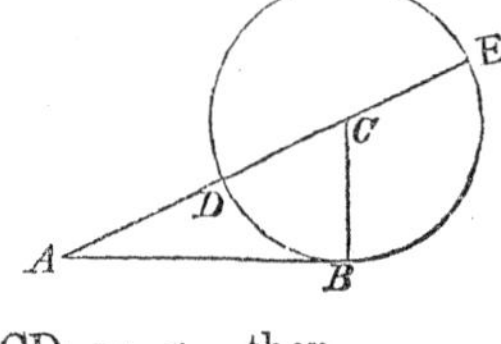

$$AC = \sqrt{b^2 + a^2}.$$

Now from C, as an origin, lay off CD $= a$, then

$$AC - CD = \sqrt{b^2 + a^2} - a = AD$$

will be represented by *the first value of x.*

From E, lay off EC $= a$, also CA $= \sqrt{b^2 + a^2}$; then

$$-EC - CA = -a - \sqrt{b^2 + a^2} = -EA$$

will be represented by *the second value of x.*

The given equation may be put under the form

$$x(x + 2a) = b^2,$$

from which we see that b is a mean proportional between x and $x + 2a$, and this relation will be satisfied by either of the above lines AD or — EA. First, by substituting AD for x, we have

$$AD(AD + 2a) = b^2 \quad \text{or} \quad AD(AD + DE) = \overline{AB}^2,$$

as it should be, since AB is a tangent, AD + DE = AE, a

secant, and AD its external part. Second, by substituting $-$ EA for x

$$-\text{EA}(-\text{ EA} + 2a) = b^2 \quad \text{or} \quad \text{EA} \times \text{AD} = \overline{\text{AB}}^2.$$

The second,

$$x^2 - 2ax = b^2,$$

gives the roots

$$x = a + \sqrt{b^2 + a^2}, \qquad x = a - \sqrt{b^2 + a^2}.$$

Construct as before, $\text{AC} = \sqrt{b^2 + a^2}$; then from C lay off $\text{CE} = a$, and

$$\text{AC} + \text{CE} = \sqrt{b^2 + a^2} + a = \text{AE},$$

will be represented by *the first value of x.*

From D, lay off $\text{DC} = a$; then from C in a contrary direction lay off $\text{CA} = \sqrt{b^2 + a^2}$, and

$$\text{DC} - \text{CA} = a - \sqrt{b^2 + a^2} = -\text{ DA}$$

will be represented by *the second value of x.*

The given equation may be put under the form

$$x(x - 2a) = b^2,$$

which will evidently be verified by the substitution of either AE or $-$ AD.

It should be observed that the values, just constructed, are the same as those for the first form, with their signs changed. This should be so, since the first form will become the second by changing x into $-x$.

The third

$$x^2 + 2ax = -b^2,$$

gives the roots

$$x = -a + \sqrt{a^2 - b^2}, \qquad x = -a - \sqrt{a^2 - b^2}.$$

From A as an origin, on the line AA′ lay off the distance $-$ AD $= -a$; at D erect the perpendicular DC $= b$; from C as a centre, with CB $= a$, as a radius, describe the arc BB′ cutting the line AA′ in B and B′; join these points with C and we shall have DB $= \sqrt{a^2 - b^2}$,

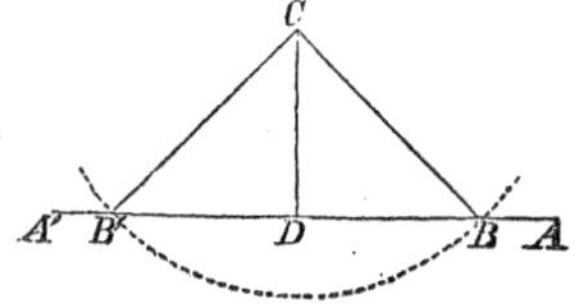

and

$$-\text{AD} + \text{DB} = -a + \sqrt{a^2 - b^2} = -\text{AB}$$

$$-\text{AD} - \text{DB} = -a - \sqrt{a^2 - b^2} = -\text{AB}'$$

will be the lines represented by the values of x.

The fourth,

$$x^2 - 2ax = -b^2,$$

gives the roots

$$x = a + \sqrt{a^2 - b^2}, \qquad x = a - \sqrt{a^2 - b^2}.$$

From A′, as an origin, lay off A′D $= a$, and make the same construction as for the third form. We thus have

$$\text{A}'\text{D} + \text{DB} = a + \sqrt{a^2 - b^2} = \text{A}'\text{B}$$

$$\text{A}'\text{D} - \text{DB}' = a - \sqrt{a^2 - b^2} = \text{A}'\text{B}'$$

for the lines represented by the values of x.

If $a = b$, both values of x reduce to $a =$ A′D. In this case, the circle does not cut the line AA′, but touches it at the point D, and the distances BD and B′D become 0. The same supposition, in the third form, reduces both values of x to $-$ AD.

If $a < b$, the values of x become imaginary in both forms; the circle neither cuts nor touches the line AA′, and the imaginary roots admit of no construction.

DETERMINATE PROBLEMS.

6. A thorough knowledge of the preceding principles, will render the solution of all determinate problems simple and easy.

Problem 1. *In a given triangle, to inscribe a square.*

Let ABC be the triangle. Represent its base, AB, by b, and its altitude CG by h. Suppose the problem to be solved, and that ODEF is the required square, its unknown side DE = EF being represented by x. Since the side DE is parallel to AB, we must have

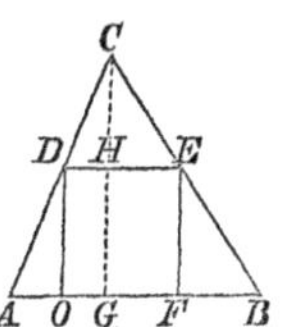

$$\text{AB} : \text{DE} :: \text{CG} : \text{CH} \qquad \text{or} \qquad b : x :: h : h - x;$$

whence

$$hx = bh - bx \qquad \text{and} \qquad x = \frac{bh}{b + h};$$

hence x is a fourth proportional to $b + h$, b and h, and may be constructed as in example 3, Art. (4). Or better thus: Produce the base AB until BL $= h$; at B and L erect the perpendiculars BN and LM; make LM $= h$ and join M and A; the part BN cut off on the first perpendicular will be represented by x. For, since BN is parallel to LM, we have

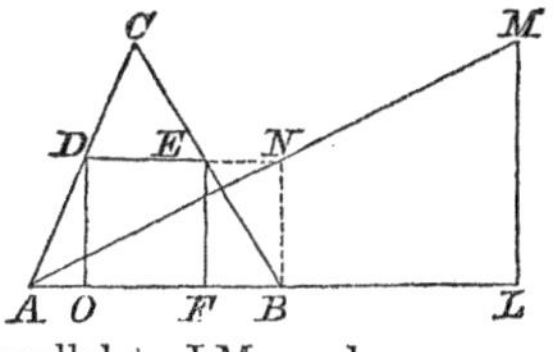

$$\text{AL} : \text{AB} :: \text{LM} : \text{BN} \qquad \text{or} \qquad b + h : b :: h : \text{BN}$$

whence

$$BN = \frac{bh}{b + h} = x.$$

Therefore, through N draw ND parallel to AB; let fall the perpendiculars EF and DO, and the square ODEF will be the required square.

The value of x, and the construction of BN, will evidently be the same for all triangles having the same base and equal altitudes. If all the angles of the triangle are acute, the square will lie wholly within the triangle as in the above figure. If there is one right angle, two sides of the square will lie upon the sides of the triangle as AD′E′F′. If there is one obtuse angle, part of the square will lie within and part without the triangle, as O′D″E″F″.

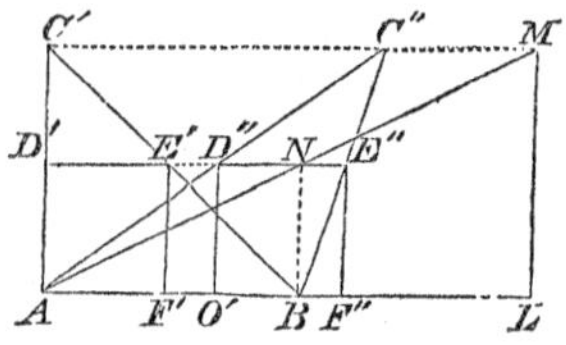

7. *Problem* 2. *In a given triangle, to inscribe a rectangle, the ratio of whose adjacent sides is known.*

Let ABC be the triangle. Let AB $= b$ and CG $= h$, and let the known ratio of the sides of the rectangle be denoted by r. Suppose the problem to be solved, and that ODEF is the required rectangle. Denote the unknown side DE, by y, and DO by x; then by the given condition, we have

$$\frac{y}{x} = r \text{............} (1).$$

Since DE is parallel to AB, we have

AB : DE : : CG : CH or $b : y :: h : h - x$

whence

$$hy = bh - bx.$$

From this, by substituting the value $y = rx$, taken from

equation (1), we deduce

$$rhx = bh - bx \quad \text{or} \quad x = \frac{bh}{b + rh}.$$

To construct this value of x; produce the base AB until BL $= rh$; through L draw LM parallel to BC until it meets CM parallel to AB, in M; join M and A; at the point E, let fall EF perpendicular to AB, it will be the required line. For, since the triangles AEB and AML are similar, their bases will be to each other as their altitudes, and we shall have

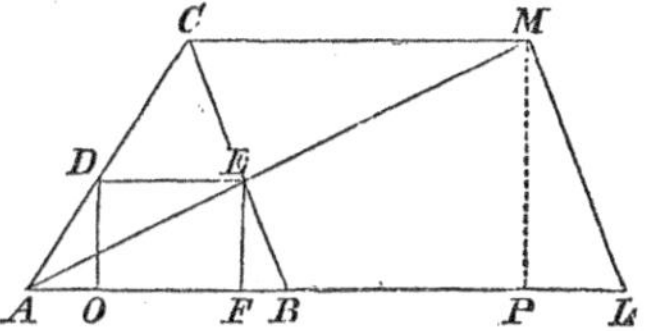

$$\text{AL} : \text{AB} :: \text{MP} : \text{EF} \quad \text{or} \quad b + rh : b :: h : \text{EF}$$

whence

$$\text{EF} = \frac{bh}{b + rh} = x.$$

Therefore, through E draw ED parallel to AB, and let fall the perpendiculars EF and DO; ODEF will be the required rectangle.

If $r = 1$, the sides are equal, the rectangle becomes a square, and we have the same value for EF as in the preceding article.

8. *Problem* 3. *To draw a straight line tangent to two given circles.*

Since the two circles are given, both in extent and position, we know their radii and the distance between their centres.

Let us denote the radius, CM, of the first circle by r, that of the second, C′M′, by r', and the distance between their centres, CC′, by a, and suppose that MM′ is the required tangent and denote the distance CT by x.

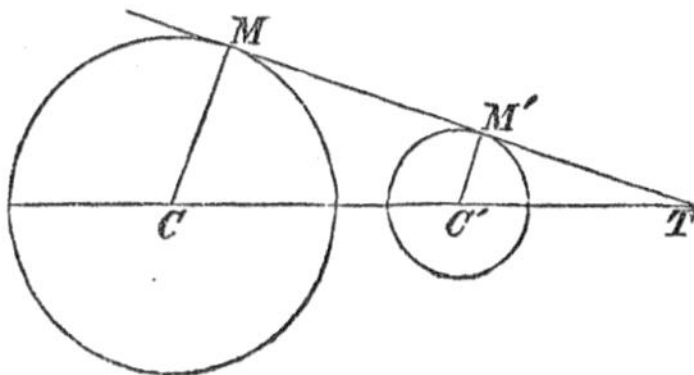

There are two cases:

First; when the tangent does not pass between the circles.

Since the radii drawn to the points of contact, M and M′, must be perpendicular to the tangent, we have CM parallel to C′M′, and hence the proportion

$$\text{CM} : \text{C'M'} :: \text{CT} : \text{C'T} \quad \text{or} \quad r : r' :: x : x - a,$$

whence

$$r'x = rx - ra \quad \text{and} \quad x = \frac{ar}{r - r'}.$$

To construct this value of x: Through the centres C and C′, draw any two parallel radii CN and C′N′, on the same side of CC′; join their extremities by the line NN′ and produce it until it meets CC′ in T; CT will be the line represented by x. For, draw N′O parallel to CC′, we then have

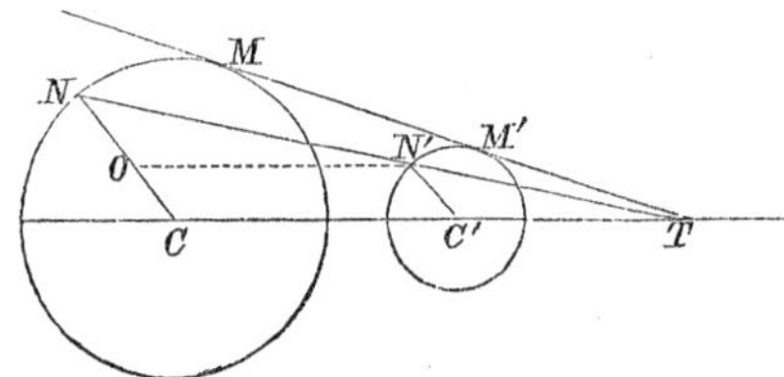

$$\text{NO} : \text{NC} :: \text{ON'} : \text{CT} \quad \text{or} \quad r - r' : r :: a : \text{CT}$$

whence

$$\text{CT} = \frac{ar}{r - r'} = x.$$

Therefore, through the point T, draw TM tangent to one of the circles, it will be tangent to the other.

If $r > r'$, the value of x is positive, and the point, T, will be on the right of C.

If $r = r'$, the two circles are equal, the value of x reduces to

$$\frac{ar}{0} = \infty;$$

the point T is at an infinite distance, and the tangent is parallel to CC′.

If $r < r'$, the value of x is negative, and the point T is on the left of C.

If $r = 0$, x will be 0, the first circle becomes a point, and the tangent is drawn from this point to the second circle.

If $r' = 0$, x will reduce to a, the second circle becomes a point, and the tangent is drawn from this point to the first circle.

If $r = 0$ and $r' = 0$, the value of x reduces to $\frac{0}{0}$, *an indeterminate quantity*, each circle becomes a point, and the tangent coincides with CC′.

Second; when the tangent passes between the circles.

In this case as in the other, the lines CM and C′M′ are parallel, hence, the triangles MCT and M′C′T are similar, and we have the proportion

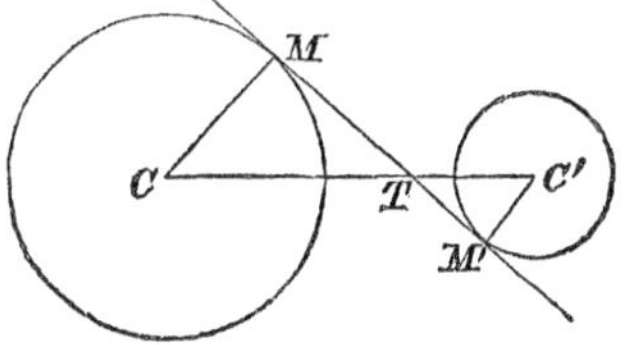

$$CM : C'M' :: CT : C'T, \quad \text{or} \quad r : r' :: x : a - x,$$

whence

$$r'x = ar - rx \quad \text{and} \quad x = \frac{ar}{r + r'}.$$

To construct this: Through C and C′ draw any two parallel radii, on different sides of CC′; join their extremities by the line NN′; CT will be the line represented by x. For, through N′, draw N′O parallel to CC′, then we have the proportion

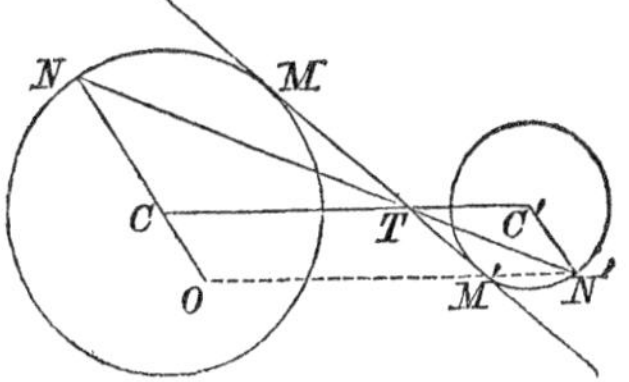

$$NO : NC :: ON' : CT, \quad \text{or} \quad r + r' : r :: a : CT,$$

whence

$$CT = \frac{ar}{r + r'} = x.$$

The value of x is positive for all values of r and r'; reduces to $\frac{a}{2}$ when $r = r'$; to 0 when $r = 0$; to a when $r' = 0$, and to $\frac{0}{0}$, when r and r' are both equal to 0.

9. *Problem* 4. *To construct a rectangle, knowing its area and the difference between its adjacent sides.*

Let a^2 denote the given area, Art. (2), and d the difference between the sides. Let x denote the least side, then $x + d$ will denote the greatest, and since the rectangle of these two sides must equal the given area, we have

$$x\ (x + d) = a^2 \qquad \text{or} \qquad x^2 + dx = a^2;$$

whence

$$x = -\frac{d}{2} \pm \sqrt{a^2 + \frac{d^2}{4}}.$$

If we take the first value

$$x = -\frac{d}{2} + \sqrt{a^2 + \frac{d^2}{4}},$$

and add d to it, we have for the greatest side

$$x + d = \frac{d}{2} + \sqrt{a^2 + \frac{d^2}{4}}.$$

To construct these values: Make $AB = a$; at B, erect the perpendicular $BC = \frac{d}{2}$, we shall have, Example 7, Art. (4),

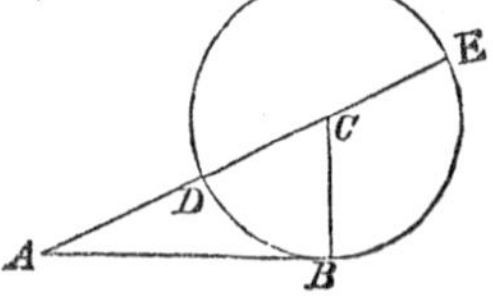

$$AC = \sqrt{a^2 + \frac{d^2}{4}}.$$

From AC, take $CD = \frac{d}{2}$, and we have

$$AD = -\frac{d}{2} + \sqrt{a^2 + \frac{d^2}{4}} = x = \textit{the least side.}$$

To AC, add $CE = \frac{d}{2}$, and we have

$$AE = \frac{d}{2} + \sqrt{a^2 + \frac{d^2}{4}} = x + d = \textit{the greatest side.}$$

and the rectangle $AE \times AD = a^2$ will be the required rectangle.

If we take the second value

$$x = -\frac{d}{2} - \sqrt{a^2 + \frac{d^2}{4}},$$

and add d to it, we have for the greatest side

$$x + d = \frac{d}{2} - \sqrt{a^2 + \frac{d^2}{4}}.$$

By examining these values, we see, that the expression for the least side, taken with a negative sign, is the same as that for the greatest side, in the first case. Also, that the expression for the greatest side, taken with a negative sign, is the same as that for the least side, in the first case. Therefore we have, in this case,

$-$ AE, *for the least side,*

$-$ AD, *for the greatest side,*

the product of which is evidently positive and equal to $\overline{AB}^2 = a^2$.

It should be observed, that it is only in an algebraic sense, that $-$ AE is less than $-$ AD, its numerical value being evidently the greatest.

It is also evident, that the two rectangles thus determined are absolutely equal, or that in reality there is but one rectangle which will fulfil the required conditions. Why then, it may be asked, do these conditions lead to an equation of the second degree? To this it may be answered, that in Algebra, properly applied, not only are problems solved in their most general sense, every possible solution being given by the equation, which is the algebraic statement of the problem; but also, whenever the conditions of a problem, expressed in two independent ways, give rise to the same equation, this equation must give an answer corresponding to each mode of expressing the conditions; that is, must be of the second degree, and it will thus be impossible to arrive at one solution disconnected from the other.

Thus, in the above problem, should we represent the greatest side by $-x$, the least would be represented by $-x-d$, and their product give

$$-x\,(-x-d) = a^2 \quad \text{or} \quad x^2 + dx = a^2,$$

the same equation found before; hence, this equation ought not only to give the least side, as at first proposed, but also another value of x, which taken with a negative sign, will represent the greatest side of the rectangle.

10. *Problem* 5. *To divide a given straight line into extreme and mean ratio.*

Let AB $= a$ be the given line. It is to be divided into two parts, such, that the greater shall be a mean proportional between the whole line and less part. Denote the greater part by x, then $a - x$ will denote the less part, and the condition will give

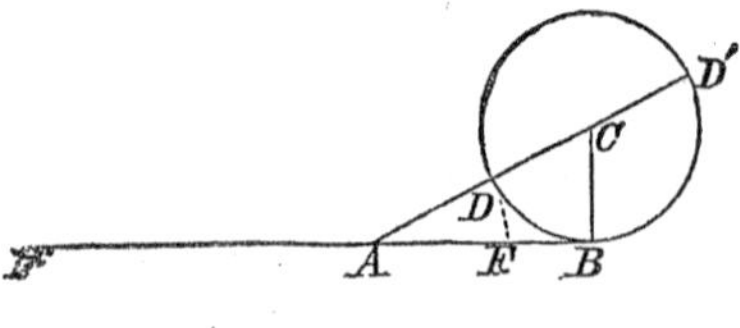

$$x^2 = a\,(a-x), \quad \text{or} \quad x^2 + ax = a^2,$$

whence

$$x = -\frac{a}{2} + \sqrt{a^2 + \frac{a^2}{4}}, \quad x = -\frac{a}{2} - \sqrt{a^2 + \frac{a^2}{4}}$$

which may be constructed precisely as in the preceding problem, the first being AD and the second $-$ AD′. With A as a centre, and AD as a radius, describe the arc DF, the line will be divided in the required ratio at F, AF being the greater part.

The second value of $x = -$ AD′ is numerically greater than AB. It can then form no part of it, and can not be an answer to the proposed question. But if we substitute it for x in the first equation, we have

$$(-\text{AD}')^2 = a\,[a - (-\text{AD}')] \quad \text{or} \quad \overline{\text{AD}'}^2 = a\,(a + \text{AD}')$$

that is, AD′ is a mean proportional between AB and AB + AD′. Since this second value of x is negative, we lay it off to the left of A, and thus construct the point F′, the distance from which to A, is a mean proportional between its distance from B and the length of the given line.

Moreover, we see that the second, as well as the first value of x, is a solution of the more general proposition, "Two points, A and B, being given, to find, on the indefinite line which joins them, a third point, the distance from which to the first shall be a mean proportional between its distance from the second and the distance between the two." To this proposition there are evidently two solutions, F on the right of A being one of the points, and F′ on its left, the other. Thus, the problem at first proposed being a particular case of a more general one, its solution, in accordance with the principle laid down in the preceding article, must necessarily draw with it that of the other case, thus giving rise to an equation of the second degree.

11. *Problem* 6. *Through a given point without a given angle,*

to draw a straight line, cutting the sides of the angle, so that the sum of the distances from the points of intersection to the vertex, shall be equal to a given line.

Let YAX be the given angle, and M the given point. Produce AX to the left, and let the two distances MP′ and MP be represented by a and b. Denote the given line by c. Suppose MN to be the required line, and denote the two unknown distances, AN by x and AO by y. Then from the condition of the problem, we have

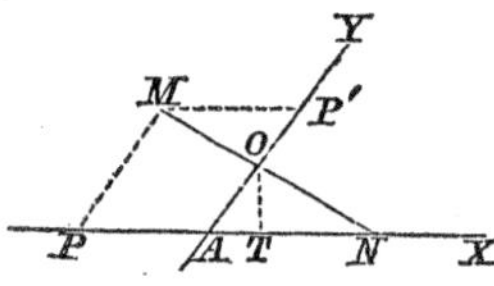

$$AN + AO = c \quad \text{or} \quad x + y = c \ldots\ldots\ldots\ldots(1).$$

But since MP is parallel to AO, we have

$$PN : AN :: PM : AO, \quad \text{or} \quad a + x : x :: b : y;$$

whence

$$y\ (a + x) = bx \ldots\ldots\ldots(2).$$

Substituting the value of x, deduced from equation (1), we have

$$y\ (a + c - y) = b\ (c - y)$$

or

$$y\ (a + c + b - y) = bc.$$

This being an equation of the second degree, its roots may be deduced and constructed as in Art. (5). But by examining it, in its present form, we see that $\sqrt{bc}$ is the ordinate of a circle whose diameter is $a + c + b$, and the corresponding segments of the diameter, y and $a + c + b - y$, which leads to a simple construction of the value of y. Thus: From P′, lay off, on AY, P′B $= c$; also BC $= a$; on AB describe the semicircle ALB; at P′ erect the ordinate P′L, it will be represented by $\sqrt{bc}$, Example 5, Art. (4). Through L draw LC′ parallel to AY; then through A and C draw the perpendiculars AA′ and CC′; A′C′ will be equal to $a + c + b$; on this line describe the semicircle C′OA′; the distance from the point O, in which it cuts AY, to A will be represented by y. For OO′ = P′L, and the segment A′O′ = AO, fulfils the required condition.

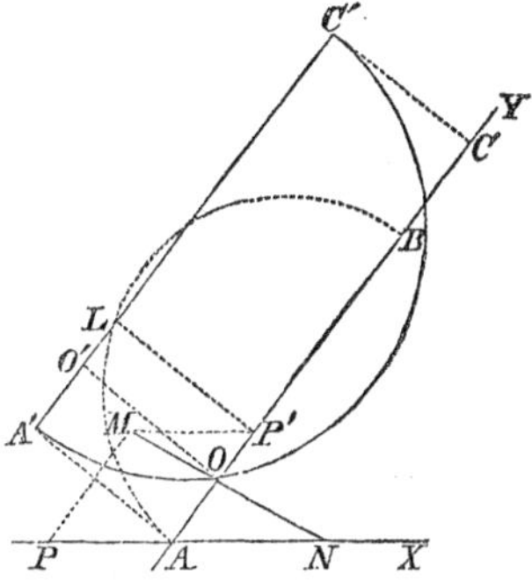

12. *Problem* 7. *Through a given point, without a given angle, to draw a straight line, so as to cut off a given area.*

Let the given point and angle be, as in the first figure of the preceding article. Let h^2 represent the given area, and β the given angle. The expression for the measure of the required triangle will be $\frac{1}{2}$OT × AN. From the right angled triangle OAT, we have

$$\text{OT} = \text{OA} \sin \text{YAX} = y \sin \beta;$$

hence the area will be expressed by

$$\tfrac{1}{2}xy \sin \beta.$$

Substituting the value of x, taken from equation (2), of the preceding article, and placing the result equal to h^2, we have

$$\frac{1}{2}\frac{ay}{b - y} y \sin \beta = h^2,$$

which, by reduction, becomes

$$y^2 + \frac{2h^2y}{a \sin \beta} = \frac{2h^2b}{a \sin \beta}.$$

Solving this equation, we obtain

$$y = -\frac{h^2}{a \sin \beta} \pm \sqrt{\frac{h^4}{a^2 \sin^2 \beta} + \frac{2h^2 b}{a \sin \beta}}.$$

To construct these values: Through A draw AA′ perpendicular to AY; then since the angle A′PA $= \beta$, we shall have

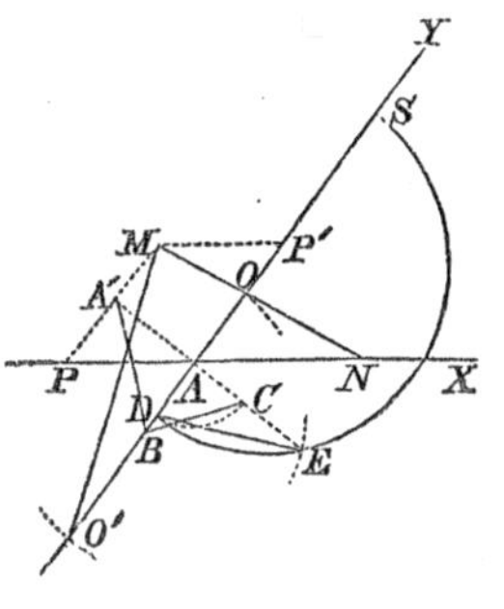

$$AA' = AP \sin \beta = a \sin \beta.$$

Upon AY, in a negative direction, lay off AB $= h$; join A′ and B, and at B erect BC perpendicular to A′B, then

$$\overline{AB}^2 = AA' \times AC$$

or

$$AC = \frac{h^2}{a \sin \beta}.$$

Since this expression is negative in the above values of y, we lay it off from A to D. The radical part of these values may be put under the form

$$\sqrt{\left(2b + \frac{h^2}{a \sin \beta}\right) \frac{h^2}{a \sin \beta}}.$$

To construct it, we lay off P′S $= b$; on SD describe a semi-circle, the chord DE will be the value of the radical, for

$$AD = \frac{h^2}{a \sin \beta}, \qquad DS = 2b + AD,$$

and DE is a mean proportional between them. From D lay off DO $=$ DE, and AO will be represented by the first value of y. From D lay off DO′ $=$ DE, and AO′ will be represented by the second value of y. Through the points O and O′, draw MO and MO′, and either triangle cut off will fulfil the condition of the problem.

13. By an examination of the manner in which the preceding problems have been solved, we may derive the following general rule for solving determinate problems.

Conceive the problem to be solved geometrically, and draw a figure containing the given and required parts, and such other lines as may be necessary to show the relation between them. Represent the known lines by the first, and the unknown by the last letters of the alphabet. Consider the geometrical relations existing between these lines, and express them by equations, taking care to deduce as many equations as there are unknown quantities. Solve these equations and construct upon a single figure the values thus deduced.

By an application of this rule the following problems are readily solved.

8. Through a given point without the circumference of a circle, to draw a straight line intersecting it, so that the chord included within, shall be equal to a given line.

9. To draw a line parallel to the base of a triangle, so as to divide it into two equal parts.

10. To inscribe, in a given triangle, a rectangle whose area is known.

11. Through two given points, to describe a circle tangent to a given right line.

PART II.

INDETERMINATE GEOMETRY.

---

14. The second branch of Analytical Geometry, which has for its principal object the analytical investigation of the general properties of lines and surfaces, is much more extended in its application, and interesting in its results, than that which we have just examined. It is called *Indeterminate Geometry*, from the fact, that, in the equations used, the unknown quantities admit of an infinite number of values, or are *indeterminate*, and are therefore called *variables*; while from the nature of the problems discussed in the first branch, they admit of a finite number of values only, and must be determinate.

OF POINTS IN A GIVEN PLANE.

15. Let AX and AY be two fixed right lines, indefinite in extent, and M any point of their plane within the angle YAX. Through this point draw MR and MP parallel respectively to AX and AY; then if the distances MR and MP are given, it is evident that the position of the point M, will be known, and may be constructed, by laying off on the line AX, beginning at A, AP = RM, drawing PM parallel to AY; then on AY, laying off

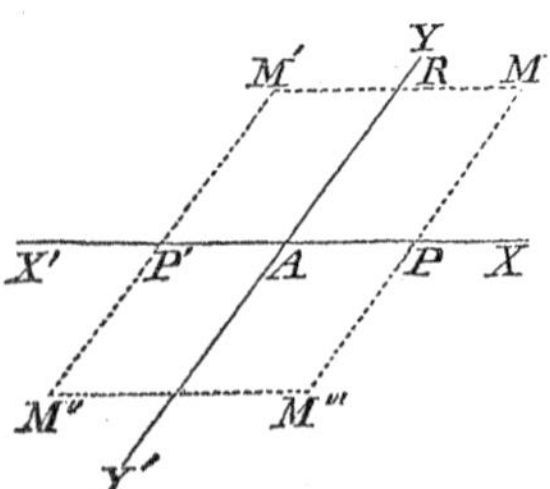

AR = PM and drawing RM parallel to AX; the point of intersection of these parallels will be the required point.

The distances MR and MP are called *the rectilineal co-ordinates* of the point. The first, or the distance of the point from AY, is *the abscissa*; and the second, or the distance of the point from AX, is the *ordinate* of the point, these distances being measured on lines parallel respectively, to AX and AY.

The fixed lines, to which the point is thus referred, are called *the axes of co-ordinates*, or *co-ordinate axes.*

Their point of intersection A, from which both abscissa and ordinate are estimated, is *the origin of co-ordinates.*

16. The abscissas of points, the position of which is indeterminate, are, in general, denoted by the letter x, and the ordinates by y, though other letters are sometimes used.

The co-ordinates of points, the position of which is known, are usually denoted either by the first letters of the alphabet, or by the symbols x', y', x'', y'', &c. If we denote the co-ordinates MR by a, and MP by b, the equations

$$x = a \qquad y = b \text{..........(1)},$$

are called *the equations of the point* M, and the values of a and b being known, the point is said to be given, and may be constructed, *in the first angle*, YAX, by laying off AP $= a$ and AR $= b$, as in the preceding article.

If, at the same time, we consider the point M′, having AP′ = AP, and P′M′ = PM, it becomes necessary to adopt some notation, by which the two points may be distinguished from each other. This notation is at once suggested, by a reference to that which is used in a similar case, for the cosine of an arc in Trigonometry, and the abscissa AP′ is regarded as *negative.* Thus the equations of a point in *the second angle*, YAX′, are

$$x = -a \qquad y = b.$$

If the point is below the axis of abscissas, its ordinate, from analogy to the sine of an arc, is regarded as negative. Thus the equations of the point M″, in *the third angle* Y′AX′, are

$$x = -a \qquad y = -b;$$

and in like manner, the equations of the point M‴, in *the fourth angle*, Y′AX, are

$$x = a \qquad y = -b.$$

Thus it appears, that by assigning proper values and signs to a and b, equations (1) may be regarded as the representatives of any point in the plane of the co-ordinate axes.

If the point is on the axis of X, (the axis of abscissas), its ordinate must be 0, and its equations

$$x = a \qquad y = 0.$$

If it is on the axis of Y, its abscissa must be 0, and its equations

$$x = 0 \qquad y = b.$$

By the essential signs of a and b, in these equations, we ascertain whether the points are on the right or left of the origin, above or below the axis of X.

If the point is on both axes at the same time, that is, at the origin, its equations, or *the equations of the origin*, become

$$x = 0 \qquad y = 0.$$

17. Let x', y', and x'', y'', be the co-ordinates of any two points, as M and M′, in the plane YAX. Join M and M′, and draw MR parallel to AX, then in the triangle MM′R we have, from Trigonometry,

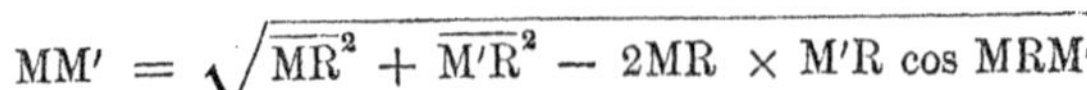

$$\text{MM}' = \sqrt{\overline{\text{MR}}^2 + \overline{\text{M}'\text{R}}^2 - 2\text{MR} \times \text{M}'\text{R} \cos \text{MRM}'}$$

the radius being supposed equal to unity. But

$$MR = PP' = x'' - x', \qquad M'R = y'' - y';$$

hence, denoting MM′ by D, the angle YAX by β, and observing that $\cos MRM' = -\cos\beta$, we have

$$D = \sqrt{(x'' - x')^2 + (y'' - y')^2 + 2(x'' - x')(y'' - y')\cos\beta} \ldots (1).$$

If $\beta = 90°$, $\cos\beta = 0$, and this formula reduces to

$$D = \sqrt{(x'' - x')^2 + (y'' - y')^2} \ldots\ldots\ldots\ldots (2),$$

that is, if the axes of co-ordinates are perpendicular to each other, *the distance between two points, in their plane, is equal to the square root of the sum of the squares of the differences of the abscissas and ordinates of the points.*

If one of the points, as M, is at the origin, x' and y' will be 0, and the last formula reduce to

$$D = \sqrt{x''^2 + y''^2}.$$

18. Let P be a *fixed point,* PS *a fixed right line,* and M any point of a plane containing PS. If the length of the line PM, which we represent by r, and the angle v, made by this line with the fixed line, are given, the position of the point will be fully determined, and may be constructed, by drawing through P a line making, with the line PS, the given angle, and then from the point P, laying off, on this line, the given distance. By varying the angle v, through all values from 0 to 360°, and the line r from 0 to infinity, the position of every point of the plane may be determined.

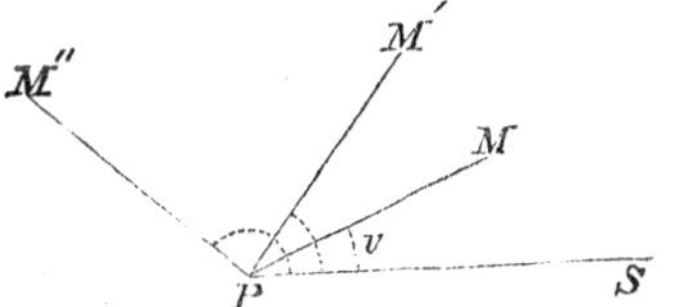

The point P is called *the pole;* the line PM, *the radius vector,* and the variables r and v, *the polar co-ordinates* of the point.

19. By a review of the preceding discussion, we see that the position of the points have been determined by ascertaining their situation with reference to certain other fixed points or magnitudes. In the first, or *system of rectilineal co-ordinates*, the points are referred to *two fixed right lines*, and the means of reference are *two other right lines*, which vary in length, as the position of the point is changed. In the second, or *system of polar co-ordinates*, the points are referred to a *fixed point* and *a fixed right line*, and the means of reference are *a variable angle* and *a variable right line.*

Although there are other methods of determining the position of points, these are the two in most general use. In every system, it should be observed, that the position, thus determined, is not absolute but relative, as all that thus becomes known, is the position of the point with reference to some other points or magnitudes; and also, that the general name of *co-ordinates of a point*, is applied *to the elements, of whatever nature, by means of which the position of the point is determined.*

OF THE RIGHT LINE IN A GIVEN PLANE.

20. Let BM be any right line, in the plane of the co-ordinate axes AX and AY, and let M be any point of the line, of which the co-ordinates AP and MP are denoted by x and y. Through the origin A, draw AM′ parallel to BM. Represent the angle YAX by β, and MBX $=$ M′AX by α; the angle PM′A $=$ M′AY will then be represented by $\beta - \alpha$.

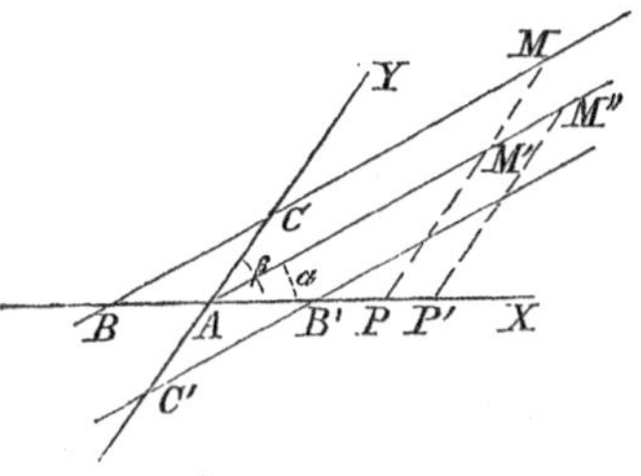

From the triangle AM′P, we have the proportion

$$AP : PM' :: \sin PM'A : \sin M'AP \quad \text{or} :: \sin(\beta - \alpha) : \sin \alpha$$

or, representing PM′ by y'

$$x : y' :: \sin(\beta - \alpha) : \sin \alpha,$$

and since AP is to PM′ as the abscissa of any other point of the line AM′ is to its corresponding ordinate, this relation will exist whatever be the position of the point M′, on the line AM′. From this proportion, we deduce

$$y' = \frac{\sin \alpha}{\sin(\beta - \alpha)} x.$$

But the ordinate of any point of the line BM, as PM, exceeds the corresponding ordinate of the line AM′, by the constant distance MM′ = AC. Representing this distance by b, we have, for every point of the line,

$$y = y' + b, \qquad \text{or} \qquad y = \frac{\sin \alpha}{\sin(\beta - \alpha)} x + b.$$

This equation expresses the relation between the co-ordinates, x and y, of every point of the line BM, and is called *the equation of the line.* The co-efficient of x, in this equation, represents the ratio of the sines of the angles which the line makes with the axes of X and Y, and the absolute term, (b), the distance from the origin to the point in which the line cuts the axis of Y.

21. By attributing, in succession, all values to α, between 0 and 360°, and all possible values to b, both positive and negative, the equation

$$y = \frac{\sin \alpha}{\sin(\beta - \alpha)} x + b \text{............} (1),$$

may be made to represent every right line in the plane of the co ordinate axes.

If b is *negative*, the line takes the position B′C′, and its equation will be

$$y = \frac{\sin \alpha}{\sin(\beta - \alpha)} x - b.$$

If $\alpha = 0$, the line is *parallel to the axis of* X, and its equation reduces to

$$y = 0.x + b,$$

or

$$y = b, \qquad x \textit{ being indeterminate.}$$

If $b = 0$, at the same time, the line coincides with the axis of X; hence, the equation of the axis of X, is

$$y = 0, \qquad x \textit{ being indeterminate,}$$

If b is *positive*, the line is above the axis of X; if *negative*, it is below it.

Solving equation (1) with respect to x, it is put under the form

$$x = \frac{\sin(\beta - \alpha)}{\sin \alpha} y - \frac{b \sin(\beta - \alpha)}{\sin \alpha} \ldots\ldots\ldots(2).$$

From the triangle BAC, we have

$$\sin \alpha : \sin(\beta - \alpha) :: b : \text{AB} = \frac{b \sin(\beta - \alpha)}{\sin \alpha},$$

that is, the absolute term of equation (2), represents the distance from the origin to the point in which the line cuts the axis of X. Representing this by a, the equation becomes

$$x = \frac{\sin(\beta - \alpha)}{\sin \alpha} y + a.$$

If in this $\alpha = \beta$, the line is *parallel to the axis of* Y, and the equation reduces to

$$x = 0.y + a,$$

or

$$x = a, \qquad y \textit{ being indeterminate.}$$

If $a = 0$, at the same time, the line coincides *with the axis of* Y, and its equation becomes

$x = 0$, *y being indeterminate.*

If a is positive, the line is on the right, if negative, it is on the left of the axis of Y.

22. Equation (1), of the preceding article, contains two kinds of quantities, x and y, which are different for different points of the line, and are therefore called *variables*; and a, β and b, which remain the same for the same line, and are called *constants*. When the values of these constants are known, the line, as we have seen, is fixed in position, or completely determined.

Since the equation contains two variables, we may assign any value to one, and, by the solution of the equation, deduce the corresponding value of the other. These two, taken together, will be the co-ordinates of a point of the line, which may be constructed as in Art. (15). By assigning other values, in succession, to one of the variables, and deducing the corresponding values of the second, any number of points may be determined, *and the line be thus constructed by points.*

Likewise, if either of the co-ordinates of a point of the line is known, the other may, at once, be deduced, by substituting the known value in the equation, and solving it with reference to the variable whose value is required. Thus, we know that the abscissa of that point of the line, which is on the axis of Y, is 0. Substituting $x = 0$, in the equation, we deduce $y = b$, which is the ordinate of the point in which the line cuts the axis of Y.

If we substitute $y = 0$, the resulting value of x, will be the abscissa of the point in which the line cuts the axis of X. This ordinate and abscissa being laid off respectively on the axes of Y and X, a right line, drawn through their extremities, will be the line to which the equation belongs.

23. From the preceding article we see that equation (1) of

Art. (21) is truly the analytical representative of the right line. And in general, any line, curved as well as straight, may be thus represented by its equation; that is, *by an equation which expresses the relation between the co-ordinates of every point of the line.*

Every such equation will contain two kinds of quantities, viz: *variables* and *constants.* The variables represent the co-ordinates of the different points of the line, and the constants serve to determine its position and extent. This is plain from the fact, that the constants being given, all the points of the line may be constructed, as in Art. (22). We therefore say, that a line is given, *when the form of its equation, and the constants, which enter it, are known.*

From the definition of the equation, it follows, *that if a point is on a given line, its co-ordinates, when substituted for the variables, must satisfy the equation of the line.*

Also, if a point is not on a given line, its co-ordinates will not satisfy the equation.

24. It will, in general, be found more convenient to take the co-ordinate axes at right angles to each other; and they will be so regarded, unless it is otherwise expressly mentioned. Under this supposition, β, in equation (1) of Art. (21), will be 90°,

$$\sin(\beta - \alpha) = \sin(90^\circ - \alpha) = \cos\alpha,$$

and the equation reduce to

$$y = \frac{\sin\alpha}{\cos\alpha}x + b, \qquad \text{or} \qquad y = \text{tang}\,\alpha x + b, \text{*}$$

or denoting tang α by a

* NOTE.—In Analytical Geometry, R or the radius of the trigonometrical tables, is always regarded as *unity*, unless it is otherwise mentioned.

$$y = ax + b \ldots\ldots\ldots\ldots (1),$$

in which, it should be remembered, that a represents the tangent of the angle which the line makes with the axis of abscissas, and b the distance from the origin to the point, in which the line cuts the axis of ordinates.

If the line passes through the origin, $b = 0$, and the equation becomes

$$y = ax.$$

If the line occupies the position of AM, the angle α being acute, a is positive, and as for all points of the line in the first angle, x is also positive, the product, ax, is positive, as it should be, since y must be positive for all points above AX. For all points in the third angle, x being negative, ax is negative, as it should be, since y must be negative for all points below AX.

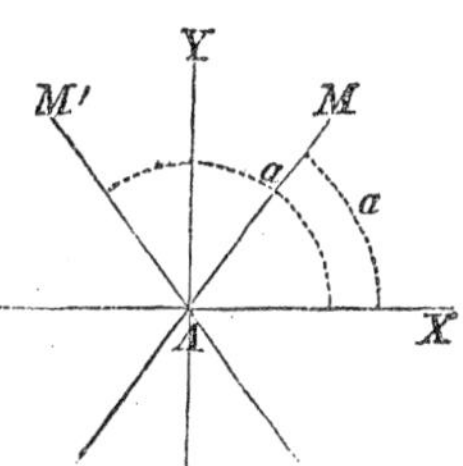

If the line occupies the position AM'; as the angle α is estimated from the axis of X, on the right of the line, around to it, as indicated in the figure; α is obtuse and a negative. For all points of the line in the second angle, x is negative and ax positive. For points in the fourth, x is positive and ax negative.

25. Every equation of the first degree, between two variables, will be a particular case of the general form

$$\mathrm{A}x + \mathrm{B}y + \mathrm{C} = 0,$$

and this, when solved with reference to y, gives

$$y = -\frac{\mathrm{A}}{\mathrm{B}}x - \frac{\mathrm{C}}{\mathrm{B}}$$

an equation of the same nature and form as

$$y = ax + b,$$

and may therefore be regarded as the equation of a right line, in which, (the axes of co-ordinates being at right angles,) $-\frac{A}{B}$ will represent the tangent of the angle which the line makes with the axis of X, and $-\frac{C}{B}$ the distance from the origin to the point in which the line cuts the axis of Y.

If the equation be solved with reference to x, it will appear under the form

$$x = a'y + b'$$

in which $a' = \frac{1}{a}$ will be the tangent of the angle made with the axis of Y, and b' the distance cut off by the line on the axis of X. Hence, *every equation of the first degree, between two variables, represents a right line; and if it be solved with reference to either variable, the coefficient of the other will be the tangent of the angle, which the line makes with the axis of that variable; and the absolute term will be the distance cut off, by the line, on the axis of that variable, with reference to which the equation is solved.*

26. The manner of constructing a right line, from its equation, may be illustrated by the following

Examples.

1. Take the equation

$$2y - 4x + 3 = 0.$$

Making $x = 0$, we deduce $y = -\frac{3}{2}$, for the ordinate of the point, in which the line cuts the axis of Y, Art. (22).

Making $y = 0$, we deduce $x = \frac{3}{4}$, for the abscissa of the point, in which the line cuts the axis of X. Assuming any convenient unit of length, and laying off $AC = -\frac{3}{2}$, and $AB = \frac{3}{4}$, BC will be the line represented by the equation.

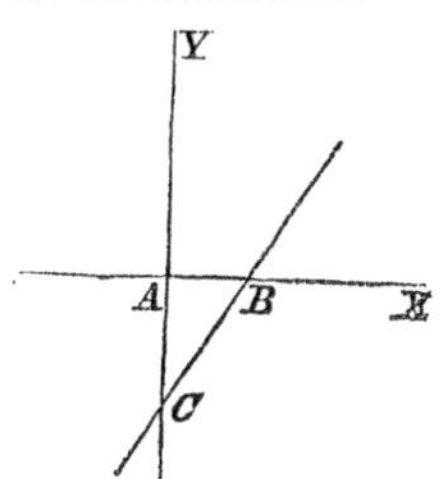

Or, thus: Solving the equation with reference to y, we have

$$y = 2x - \frac{3}{2}.$$

Laying off the distance $AC = -\frac{3}{2}$, and drawing the line CB, making, with the axis of X, an angle whose tangent is 2,* it will be the line.

Or the line may be constructed by points, thus: Making

$$x = 1 \quad \text{we deduce} \quad y = \frac{1}{2},$$

$$x = 2 \quad \text{"} \quad \text{"} \quad y = \frac{5}{2},$$

&c.

The points, represented by the different sets of co-ordinates thus determined, may be constructed and the line drawn through them.

2. $$3y + 9x - 1 = 0.$$

---

* NOTE.—An angle whose tangent is a given number may always be constructed thus. Let $\text{tang } a = \frac{c}{d}$. Lay off $AB = d$ and erect the perpendicular $BM = c$; draw AM, the angle MAB will be the required angle. For we have

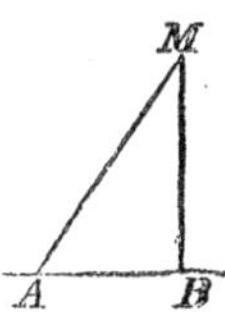

$$\text{tang MAB} = \frac{BM}{AB} = \frac{c}{d} = \text{tang } a.$$

When tang a is a whole number, as in the example, $d = AB = 1 =$ *the unit of length.*

3. $$y - x - 4 = 0.$$

4. $$2y + 3x + 5 = 0.$$

27. Let
$$y = ax + b$$
$$y = a'x + b'$$
be the equations of two right lines. Those values of x and y which, taken together, will satisfy both of these equations, must be the co-ordinates of a point on each line, Art. (23). But if we combine the two equations and deduce the values of x and y, we obtain all which can possibly satisfy both equations at the same time; these values must then be the co-ordinates of all points common to the lines.

Placing the second members of the equations equal, we have
$$ax + b = a'x + b',$$
whence
$$x = \frac{b' - b}{a - a'}.$$

Substituting this value of x, in the first equation, we obtain
$$y = \frac{ab' - a'b}{a - a'}.$$

These values of x and y must be the co-ordinates of a point common to both lines. And, in general, since the equations of right lines are of the first degree, the values resulting from their combination must be real and give one common point, and only one.

If $a = a'$, the values of x and y, both reduce to infinity; the point of intersection is then at an infinite distance, that is, *the lines are parallel.*

If $b = b'$, at the same time, both values become $\frac{0}{0}$, or *indeterminate*, as they should, since in this case the two lines coincide and have all their points common.

It is evident that the above reasoning will apply to any lines, straight or curved, and we may therefore give the following rule for obtaining the points of intersection of any two lines. *Combine the equations of the lines, and deduce the values of the variables. For each couple of real values there will be a common point. If the values are all imaginary, there will be no common point.*

Find the point of intersection of the two right lines given by the equations

$$3y - 2x + 1 = 0, \qquad 5y + 3x = 0.$$

28. Let

$$y = ax + b$$

$$y = a'x + b'$$

be the equations of any two given right lines, making the angles α and α', respectively, with the axis of X, and the angle V with each other. By the figure, we see that

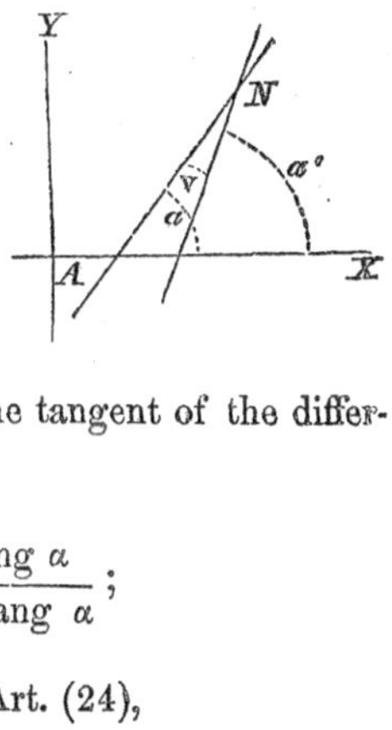

$$\alpha' = V + \alpha, \quad \text{or} \quad V = \alpha' - \alpha,$$

and by the trigonometrical formula, for the tangent of the difference of two angles,

$$\text{tang } V = \frac{\text{tang } \alpha' - \text{tang } \alpha}{1 + \text{tang } \alpha' \text{ tang } \alpha};$$

and since from the equations of the lines, Art. (24),

$$a = \text{tang } \alpha \qquad a' = \text{tang } \alpha',$$

we have

$$\text{tang } V = \frac{a' - a}{1 + a'a};$$

from which, by the substitution of the values of a' and a, given by particular equations, the natural tangent of the angle between two right ines may be found; and from a table of natural sines, cosines, &c., the value of the angle, in degrees, minutes, &c., may be determined.

If $a = a'$,

$$\text{tang } V = \frac{0}{1 + a'a} = 0;$$

hence, in this case, the angle is 0, and *the lines are parallel*, as shown in the preceding article.

If $1 + aa' = 0$,

$$\text{tang } V = \frac{a' - a}{0} = \infty,$$

the angle is 90°, and the two lines are *perpendicular to each other*.

To ascertain then, practically, whether two right lines are parallel or perpendicular: *Solve their equations with reference to either variable; if the coefficients of the other variable are equal, the lines are parallel; if the product of these coefficients plus unity is equal to 0, they are perpendicular.*

Apply this rule to the equations

1. $2y - 4x + 7 = 0, \qquad y - 2x - 3 = 0.$

2. $y - 3x + 1 = 0, \qquad 6y + 2x - 5 = 0.$

29. Let x' and y' be the co-ordinates of a given point, and

$$y = ax + b \text{............} (1),$$

the general equation of a right line, in which a and b are undetermined. If the given point is on the line, its co-ordinates, when substituted for x and y, must satisfy the equation, Art. (23), and we must have

$$y' = ax' + b,$$

an equation expressing the condition that the point, (x', y'), shall be on the line. This condition may be introduced into equation (1) by subtracting it, member from member. We thus obtain

$$y - y' = a(x - x') \ldots\ldots\ldots\ldots (2),$$

which is the equation of a right line with the condition introduced, that a given point shall be on it; or, is *the equation of a right line passing through a given point.*

a remains undetermined, as it should, since an infinite number of right lines may be drawn through the given point. If the abscissa and ordinate of the given point are 2 and 3, the equation of the line becomes

$$y - 3 = a(x - 2).$$

30. If the line, represented by equation (2) of the preceding article, be subjected to the condition that it shall be parallel to a given line, as the one whose equation is,

$$y = a'x + b,$$

we must have, Art. (28),

$$a = a'.$$

Substituting this known value in equation (2), the line will be fixed and its equation become

$$y - y' = a'(x - x').$$

If the line is required to be perpendicular to the given line, we must have

$$1 + aa' = 0, \qquad \text{or} \qquad a = -\frac{1}{a'},$$

and the equation becomes

$$y - y' = -\frac{1}{a'}(x - x').$$

1. Find the equation of a right line, passing through the point, $x' = -2$, $y' = 3$, and parallel to the line whose equation is

$$2y - x + 2 = 0.$$

2. Find the equation of a right line, passing through the same point and perpendicular to the same line.

31. If the line represented by equation (2), Art. (29), be subjected to the condition, that it shall pass through another point whose co-ordinates are x'' and y'', these co-ordinates must satisfy the equation and give the equation of condition

$$y'' - y' = a(x'' - x'),$$

from which, the value of a becomes known, and we have

$$a = \frac{y'' - y'}{x'' - x'}.$$

Substituting this, in equation (2), we obtain

$$y - y' = \frac{y'' - y'}{x'' - x'}(x - x')\text{.............}(1),$$

for *the equation of a right line passing through two given points.*

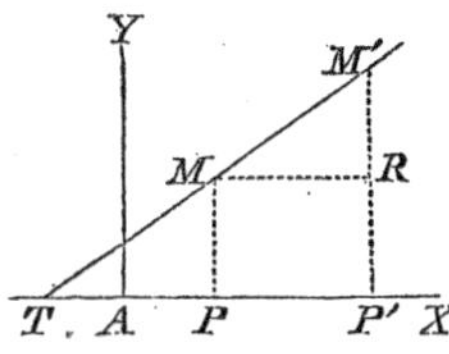

If M and M′ are the points, the co-ordinates of the first being x', y', and of the second x'', y'', we have

$$M'R = y'' - y', \quad MR = x'' - x',$$

$$\text{tang } M'MR = \text{tang } M'TX = \frac{M'R}{MR} = \frac{y'' - y'}{x'' - x'} = a.$$

If $y'' = y'$, this value of a reduces to

$$a = \frac{0}{x'' - x'} = 0,$$

as it should, since the line becomes parallel to the axis of X.

If $x'' = x'$,

$$a = \frac{y'' - y'}{0} = \infty,$$

as in this case the line is perpendicular to the axis of X.

If $x'' = x'$ and $y'' = y'$,

$$a = \frac{0}{0} \textit{ indeterminate},$$

since the two points become one, through which an infinite number of right lines may be drawn.

1. If the co-ordinates of the points are $x' = 2$, $y' = -1$; $x'' = 3$, $y'' = 0$; equation (1) will become

$$y + 1 = \frac{1}{1}(x - 2),$$

which reduces to

$$y = x - 3.$$

2. Find the equation of a right line passing through the two points

$$x' = -1, \quad y' = -2; \qquad x'' = 4, \quad y'' = -5$$

32. In every equation containing but two variables, we may, as in Art. (22), assign to one a series of values, in succession, and deduce the corresponding values of the other, and thus construct a series of points, which being joined, will evidently form a line, which will be represented by the given equation. Hence we

say, in general, *that every equation between two variables, is the equation of a line, either straight or curved.*

If all values of the first variable give imaginary values for the second, the line is said to be imaginary.

If there is but a limited number of couples of real values, which will satisfy the equation, it will represent a point or a limited number of distinct points.

33. Whenever the relation between the co-ordinates of the points of a line can be expressed by the ordinary operations of Algebra, that is, by addition, subtraction, multiplication, division, the formation of powers denoted by constant exponents, or the extraction of roots indicated by constant indices, the line is said to be *Algebraic.*

When this relation can not be so expressed, the line is *Transcendental.*

Algebraic lines only, will be considered in this Treatise. They are classed into *orders*, according to *the degree of their equations.* Thus, *a line of the first order*, is one whose equation is of *the first degree. A line of the second order*, one whose equation is of *the second degree*, &c. We have seen, Art. (25), that the right line is the only line of the first order.

The discussion of the equation of a line consists in classing the line, determining its form, its limits, its position with respect to the co-ordinate axes, and the points in which it cuts these axes.

OF THE CIRCLE.

34. Let x' and y' be the co-ordinates of the centre of a circle, and R its radius, and let x and y be the co-ordinates of any point of its circumference. The distance from the centre to any point of the circumference, will then, Art. (17), be denoted by

$$\sqrt{(x - x')^2 + (y - y')^2};$$

but, from the definition of a circumference, this distance must be constantly equal to the radius, R; hence we have

$$\sqrt{(x - x')^2 + (y - y')^2} = R,$$

or

$$(x - x')^2 + (y - y')^2 = R^2 \ldots\ldots\ldots\ldots(1);$$

and since this expresses the relation between the co-ordinates of every point of the circumference, Art. (23), it is the equation of the circumference, or *the equation of the circle*; the word circle being commonly used for circumference.

The circle will be given, when x', y', and R are given, Art. (23), and by attributing different values to these constants, we may place the centre in any position, and give to the circle any extent.

For those points of the circle which lie on the axis of X, $y = 0$; substituting this in equation (1), the corresponding values,

$$x = x' \pm \sqrt{R^2 - y'^2},$$

will be the abscissas of the points, in which the circle cuts the axis of X.

If $y' < R$, these values will be real, and the circle will intersect the axis, in two points.

If $y' = R$, the two points will unite, and the circle will be tangent to the axis of X.

If $y' > R$, the values of x will be imaginary, and there will be no point of intersection.

Each position of the circle is shown, in the accompanying figure.

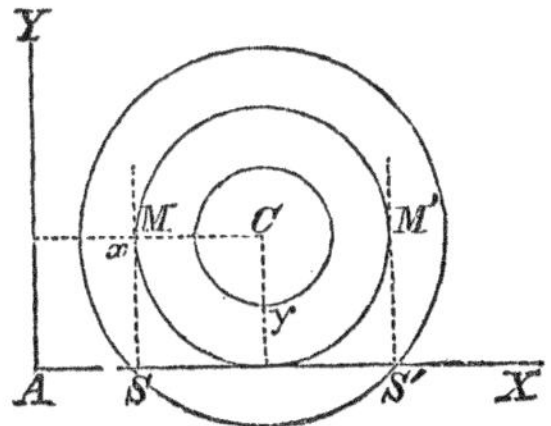

By making $x = 0$, we deduce

$$y = y' \pm \sqrt{R^2 - x'^2},$$

for the ordinates of the points, in which the circle intersects the axis of Y, and these will be real, equal, or imaginary, according as x is less, equal to, or greater than R.

Solving equation (1) with reference to y, we have

$$y = y' \pm \sqrt{R^2 - (x - x')^2}.$$

By assigning values to x, in succession, we deduce the corresponding values of y, and thus determine as many points of the curve as we please, Art. (32).

Every value of x, which makes $(x - x')^2 < R^2$, will give two real values of y. For every such value, there will, consequently, be two corresponding points of the curve.

If $x = x' + R$ or $x' - R$, the values of y will be $y' \pm 0$; the two points will unite, and the corresponding ordinate will be tangent to the curve, as SM or S′M′.

If $x > x' + R$ or $< x' - R$, the values of y will be imaginary, and there will be no corresponding points of the curve.

We thus see that the curve is limited, in the direction of the axis of X, by the two lines, SM and S′M′. In the same way, by solving the equation with reference to x, we may obtain the limits in the direction of the axis of Y.

35. If x' and y' are both equal to 0, the centre of the circle will be at the origin of co-ordinates, and equation (1), of the preceding article, will reduce to

$$x^2 + y^2 = R^2 \ldots\ldots\ldots\ldots(1).$$

To discuss this equation, Art. (33); make $y = 0$, we thus obtain

$$x = \pm R,$$

which shows, that the curve cuts the axis of X, in the two points, B and C, at distances, on the right and left of the origin, each equal to R.

Making $x = 0$, we obtain

$$y = \pm R,$$

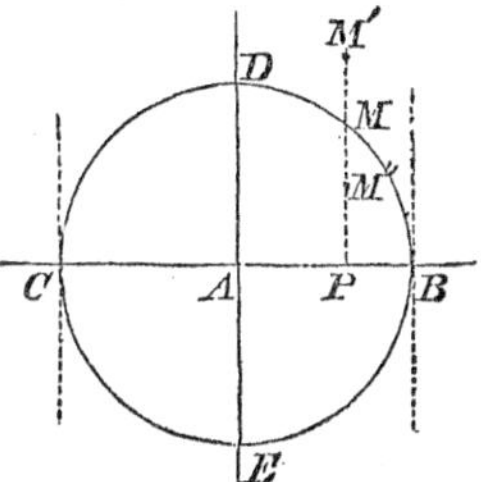

which shows that the curve cuts the axis of Y, in the two points D and E.

Solving the equation with reference to y, we have

$$y = \pm \sqrt{R^2 - x^2},$$

from which, we see that every value of x, positive or negative, and numerically less than R, gives two real values of y, equal with contrary signs; hence, for each of these values there are two corresponding points, one above, and the other below the axis of X, at equal distances from it, and the ordinates of these points, taken together, form a chord, which is bisected by the axis of X. This proves that the curve is symmetrical with respect to the axis of X.

If $x = \pm R$, y becomes equal to ± 0, which proves that the corresponding ordinates, produced, are tangent to the curve.

If x is numerically greater than R, either positive or negative, the values of y are imaginary, and there are no corresponding points of the curve. The curve is therefore limited in the direction of the axis of X, by the two tangents at B and C.

In a similar way, it may be proved, that the curve is symmetrical with respect to the axis of Y, and that its limits are two tangents at D and E.

36. For every point of the curve, as M, in the figure of the preceding article, we have

$$y^2 = R^2 - x^2 = (R + x)(R - x) = CP \times PB,$$

a well known property of the circle.

37. If y represents the ordinate of any point, as M′, without the circle, in the figure of Art. (35), we have

$$M'P > MP \quad \text{or} \quad y^2 > R^2 - x^2.$$

For any point, as M″, within the circle, we have

$$M''P < MP \quad \text{or} \quad y^2 < R^2 - x^2.$$

Hence we deduce the three analytical conditions

$y^2 + x^2 - R^2 = 0$, *for a point on the circle,*

$y^2 + x^2 - R^2 > 0$, " " *without the circle,*

$y^2 + x^2 - R^2 < 0$, " " *within the circle.*

38. If the origin of co-ordinates is at C, in the figure of Art. (35), the co-ordinates of the centre will be

$$x' = R \qquad y' = 0,$$

and the general equation (1), Art. (34), will reduce to

$$x^2 + y^2 - 2Rx = 0, \quad \text{or} \quad y^2 = 2Rx - x^2.$$

This equation has no absolute term, or term independent of x and y; the substitution of $x = 0$ and $y = 0$, will therefore satisfy it, which verifies the fact, that the origin of co-ordinates is on the curve; and, in general, *if the equation of a line has no absolute term, the line passes through the origin of co-ordinates.*

OF POINTS IN SPACE.

39. By space is to be understood, that infinite extent, in which all bodies are situated. As the absolute places of points and magnitudes, in this indefinite space, can not be determined, we have only to seek their situation, with reference to certain other objects,

which do not change their position with respect to each other. In a plane, we have seen that the situation of points and lines, is thus determined by a reference to two fixed objects, Art. (19). In space it is found necessary to refer them to three, the means of reference, as before, being called the co-ordinates.

40. Let XAY, XAZ, and YAZ, be any three fixed planes, indefinite in extent, intersecting each other in the lines, AX, AY and AZ; and let M be any point within the solid angle formed by these planes. Through this point, draw the lines MP, MP′ and MP″, respectively parallel to the lines, AZ, AY and AX, terminating in the planes. If the distances MP, MP′ and MP″, or their equals, AR, AR′ and AR″ are given, it is evident that the position of the point will be fully determined, and may be constructed, thus: On AX lay off AR″ = MP″; on AY lay off AR′ = MP′; through their extremities draw the lines R″P and R′P parallel respectively to AY and AX; through their point of intersection, P, draw PM parallel to AZ, and on it lay off the given distance, MP; the extremity will be the required point.

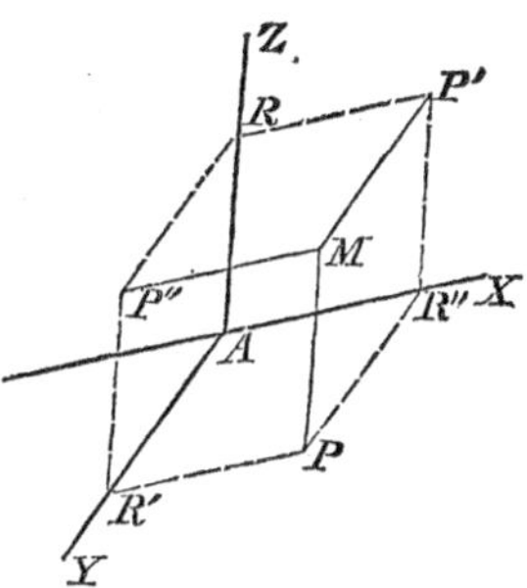

The planes XAY, XAZ and YAZ, are called *the co-ordinate planes.*

The first is designated as the plane XY; the second, as XZ; and the third, as YZ.

The lines AX, AY and AZ, are *the co-ordinate axes.*

The first is *the axis of* X, and the distances parallel to it are denoted by x. The second is *the axis of* Y, and the distances parallel to it are denoted by y. The third is *the axis of* Z, and the corresponding distances are denoted by z. The point A is *the origin of co-ordinates*, and the distances MP, MP′ and MP″, are *the rectilineal co-ordinates* of the point M.

41. If the distances of a point, from the co-ordinate planes, YAZ, XAZ and XAY, are respectively denoted by a, b and c, we have for this point

$$x = a \qquad y = b \qquad z = c,$$

which are *the equations of the point*; and when these equations are given, the point is said to be given, and may be constructed as in the preceding article.

The point M is in *the first angle*, that is, in the angle to the right of YZ, in front of XZ, and above XY.

Those points which are on the left of the plane YZ, are distinguished from those on the right, by giving the minus sign to x; those behind the plane XZ, from those in front, by giving the minus sign to y; and those below the plane XY from those above, by giving the minus sign to z. Thus, for a point in *the second angle*, that is, in the angle to the left of YZ, in front of XZ, and above XY, the equations are

$$x = -a \qquad y = b \qquad z = c.$$

For a point in *the third angle*, which is immediately behind the second,

$$x = -a \qquad y = -b \qquad z = c.$$

For a point in *the fourth angle*, immediately behind the first,

$$x = a \qquad y = -b \qquad z = c.$$

For a point in *the fifth angle*, under the first,

$$x = a \qquad y = b \qquad z = -c.$$

For a point in *the sixth angle*, under the second,

$$x = -a \qquad y = b \qquad z = -c.$$

For a point in *the seventh angle*, under the third,

$$x = -a \qquad y = -b \qquad z = -c.$$

For a point in *the eighth angle*, under the fourth,

$$x = a \qquad y = -b \qquad z = -c.$$

If a point is in the plane XY, the value of z for this point is 0, and the equations of a point, in this plane, are

$$x = a \qquad y = b \qquad z = 0;$$

and there are similar equations for points in each of the other co ordinate planes.

If a point is on the axis of X, the values of y and z, for this point, are both 0, and the equations of a point on this axis are

$$x = a \qquad y = 0 \qquad z = 0;$$

and there are similar equations for points on each of the other co-ordinate axes.

The equations of the origin of co-ordinates are

$$x = 0 \qquad y = 0 \qquad z = 0.$$

42. It is found most convenient, in practice, to take the co-ordinate planes at right angles to each other, and they are always considered to be in this position, unless it is otherwise indicated.

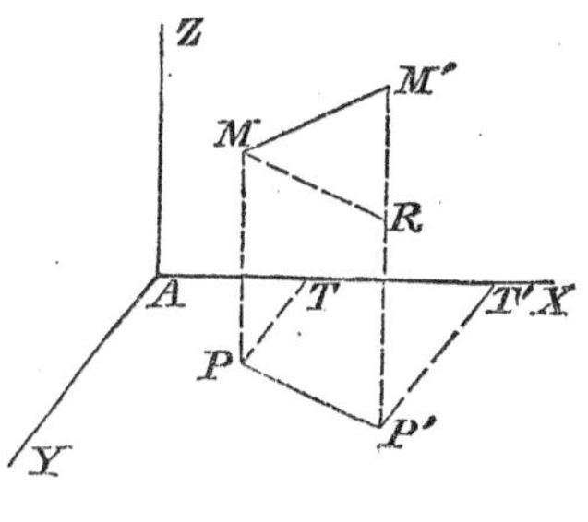

Let x', y', z', and x'', y'', z'', be the co-ordinates of any two points in space, as M and M'. Then

$$x' = \text{AT}, \; y' = \text{TP}, \; z' = \text{MP}.$$

$$x'' = \text{AT}', \; y'' = \text{T}'\text{P}', \; z'' = \text{M}'\text{P}'.$$

Join P and P', and draw

MR parallel to PP′. Then from the triangle MRM′, right angled at R, we have

$$\text{MM}' = \sqrt{\overline{\text{MR}}^2 + \overline{\text{M}'\text{R}}^2}.$$

But, Art. (17),

$$\overline{\text{MR}}^2 = \overline{\text{PP}'}^2 = (x'' - x')^2 + (y'' - y')^2,$$

and

$$\overline{\text{M}'\text{R}}^2 = (z'' - z')^2;$$

hence, denoting the distance MM′ by D, we obtain

$$\text{D} = \sqrt{(x'' - x')^2 + (y'' - y')^2 + (z'' - z')^2};$$

or, *the distance between two points, in space, is equal to the square root of the sum of the squares of the differences of their co-ordinates.*

If one of the points, as M, be placed at the origin, x', y' and z' become 0, and

$$\text{D} = \sqrt{x''^2 + y''^2 + z''^2}.$$

43. The position of points, in space, may also be determined by referring them to any other three fixed objects. For instance, let A be a fixed point, and AX a fixed line in the given plane YAX, and let M be any point in space. If the distance AM, and the angles MAP and PAX are given, the position of the point is known, and may readily be constructed.

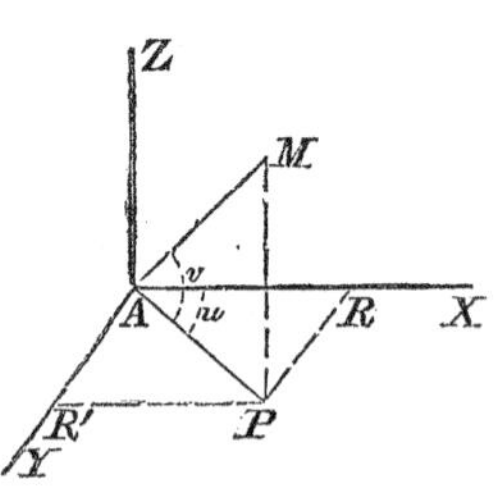

This method, in which points are referred to *a fixed point, a fixed plane,* and *a fixed line* of the plane, is called *the system of*

polar co-ordinates in space; in which the point A, is *the pole*, and the distance AM, *the radius vector*. The three variable co-ordinates are, the radius vector, the angle which it makes with the plane, and the angle which its projection on the plane makes with the fixed line.

This, and the method of rectilineal co-ordinates, discussed in the preceding article, form the two principal systems of co-ordinates in space.

OF THE RIGHT LINE IN SPACE.

44. Let

$$x = az + \alpha$$

be the equation of a right line, B′C, in the co-ordinate plane XZ, and

$$y = bz + \beta$$

the equation of B″C′, in the plane YZ. If through each of these lines, a plane be passed perpendicular to the planes XZ and YZ respectively, these planes will intersect in a right line, BC, which will thus be completely determined. The two equations

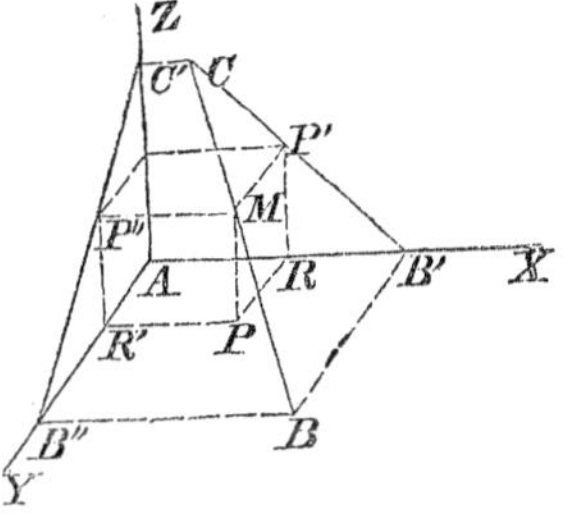

$$x = az + \alpha \ldots\ldots\ldots(1), \qquad y = bz + \beta \ldots\ldots\ldots(2),$$

taken together, may then be regarded as the equations of the right line in space, and when they are given, the right line will be given, and may be constructed by points. For, if a value be assigned to either variable, in these equations, the values of the other two can at once be deduced, and the three, taken together, will be the co ordinates of a point of the line. For instance, assume a value for z = RP′; this, with the corresponding value of x deduced from equation (1), will determine a point, P′, on the line B′C, through which if a perpendicular, P′M, be drawn to the plane XZ, it will intersect the given line in a point, M. This same value of z, with the corresponding value of y, deduced from equation (2), will determine a point, P″, on B″C′, through which, if a perpendicular be drawn to YZ, it will intersect the line, in space, at the same point, M, since no two points of this line can have the same value of z.

The two planes passing through the line in space, perpendicular to the co-ordinate planes, are called *the projecting planes of the line*; and the lines B′C and B″C′, in which they intersect the co-ordinate planes, are *the projections of the given line.*

In equation (1), a represents the tangent of the angle which the projection of the given line, on the plane XZ, makes with the axis of Z, and α the distance cut from the axis of X, by the same projection, Art. (25).

In equation (2), b represents the tangent of the angle which the projection on YZ, makes with the axis of Z, and β, the distance cut from the axis of Y.

If we combine equations (1) and (2), and eliminate the variable z, we deduce

$$y - \beta = \frac{b}{a}(x - \alpha)\ldots\ldots\ldots\ldots(3),$$

which, expressing the relation between y and x for points of the line, is evidently the equation of its projection on the plane YX.

45. The principle that the constants in the equation of a line, serve to determine it, Art. (23), may be well illustrated by supposing the four constants in equations (1) and (2) of the preceding article, to be given in succession. Thus, if a alone is given, the line is subjected to the single condition, that its projection on the plane XZ, shall make a given angle with the axis of Z, that is, it

may lie in either one of a system of parallel planes, perpendicular to XZ, and making, with the axis of Z, an angle, the tangent of which is a : If α is now given, the distance cut off on the axis of X is known, and the line may have any position in one of the before described planes : If b is also given, the other projection must make a given angle with the axis of Z, that is, the line in this fixed plane must make an angle with the axis of Z, the tangent of which is b, or it may occupy any one of an infinite number of parallel positions in this plane : If β is also given, the line is absolutely fixed.

If α and β are 0, the line will pass through the origin of co-ordinates, and its equations become

$$x = az, \qquad y = bz \ldots\ldots\ldots\ldots(1).$$

If in these,

$$a = 0, \qquad \text{and} \qquad b = 0,$$

the line will *coincide with the axis of* Z, and the equations become

$$x = 0, \qquad y = 0, \qquad z \quad \textit{indeterminate}.$$

If the value of z be taken from the first of equations (1), and substituted in the second, we obtain

$$z = \frac{1}{a}x, \qquad y = \frac{b}{a}x,$$

for the equations of the projections of the right line, passing through the origin, on the planes ZX and YX. If in these,

$$\frac{1}{a} = 0 \qquad \text{and} \qquad \frac{b}{a} = 0,$$

the line will *coincide with the axis of* X, and the equations of this axis be

$$z = 0, \qquad y = 0, \qquad x \quad \textit{indeterminate}.$$

In a similar way, if the line coincide with the axis of Y, we have

$$\frac{1}{b} = 0 \qquad \text{and} \qquad \frac{a}{b} = 0,$$

and the equations of this axis will be

$$z = 0, \qquad x = 0, \qquad y \text{ indeterminate.}$$

46. For the point in which a line pierces the plane XY, z must be 0. Substituting this value in equations (1) and (2) of Art. (44), we have

$$x = \alpha, \qquad y = \beta;$$

hence, α and β, taken together, are the co-ordinates of the points in which the right line pierces the plane XY.

In a similar way, the co-ordinates of the points in which the line pierces the other co-ordinate planes, may be determined.

47. Let

$$x = az + \alpha \text{.........}(1), \qquad y = bz + \beta \text{...........}(2),$$

$$x = a'z + \alpha' \text{.........}(3), \qquad y = b'z + \beta' \text{.........}(4),$$

be the equations of two right lines. If these lines intersect, or have a point in common, the co-ordinates of this point must satisfy the equations at the same time; or for this point, x, y and z must be the same in all of the equations. Hence, if we combine these equations and find proper values for x, y and z, they will be the co-ordinates of the common point. These four equations, containing but three unknown quantities, can not be satisfied by the same set of values if they are independent of each other. If the lines intersect, there must then be such a relation existing between the known quantities of the equations, as to make one dependent upon the other three, and the equation which expresses this relation will be the equation of condition that the lines shall intersect.

Equating the second members of (1) and (3), we deduce

$$z = \frac{\alpha' - \alpha}{a - a'},$$

and in a similar way, from (2) and (4),

$$z = \frac{\beta' - \beta}{b - b'},$$

Placing these values equal to each other, we have

$$\frac{\alpha' - \alpha}{a - a'} = \frac{\beta' - \beta}{b - b'},$$

or

$$(\alpha' - \alpha)(b - b') = (\beta' - \beta)(a - a') \text{............} (5),$$

for *the equation of condition that the lines shall intersect.*

This equation contains eight arbitrary constants, any seven of which may be assumed at pleasure, and the remaining one thus determined, so as to cause the lines to intersect.

Substituting the first of the above values of z in equation (1), and the second in equation (2), we find

$$x = \frac{a\alpha' - a'\alpha}{a - a'} \qquad y = \frac{b\beta' - b'\beta}{b - b'}.$$

These values of x and y, with either value of z, will give a point of intersection when equation (5) is satisfied.

If $a = a'$ and $b = b'$, equation (5) is satisfied, and the values of x, y and z become infinite. The point of intersection is then at an infinite distance, that is, *the lines are parallel.*

$$a = a' \qquad b = b'$$

are then the analytical conditions that two right lines, in space, shall be parallel. But $a = a'$ is the condition that the lines represented by equations (1) and (3) shall be parallel, Art. (28), and $b = b'$, the condition that the lines represented by (2) and (4) shall be parallel. Hence, *if two right lines, in space, are parallel, their projections on the same co-ordinate plane will be parallel.*

If at the same time $\alpha = \alpha'$ and $\beta = \beta'$, the above values of z, x and y become indeterminate, as they should, since the two lines then coincide.

48. Since the angle included between two right lines, in space, is the same as that included between two lines passing through a common point and parallel respectively to the first; let the lines AP and AP′ be drawn through the origin of co-ordinates, parallel to any two given lines, making with each other an angle denoted by V. The equations of AP and AP′ will be

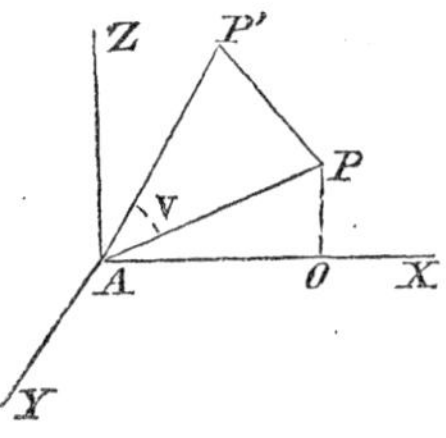

$$x = az, \qquad y = bz;$$

$$x = a'z, \qquad y = b'z;$$

in which a, b, a' and b', are the same as in the equations of the given lines, Art. (44), and the included angle is equal to V. Denote the angles, made by the first line with the axes of X, Y and Z respectively, by X′, Y′ and Z′, and let X″, Y″ and Z″ represent the corresponding angles made by the second line.

Take any point, as P, of the first line, and denote its co-ordinates by x', y' and z', and its distance from A, by r', and let x'', y'' and z'', be the co-ordinates of any point, as P′, of the second line, and r'' its distance from A, and let D be the distance PP′. Then from Trigonometry, we have

$$\cos V = \frac{r'^2 + r''^2 - D^2}{2r'r''},$$

or

$$D^2 - r'^2 - r''^2 + 2r'r'' \cos V = 0 \ldots\ldots\ldots\ldots(1),$$

in which, Art. (42),

$$D^2 = (x' - x'')^2 + (y' - y'')^2 + (z' - z'')^2 \ldots\ldots\ldots\ldots(2).$$

But if from P, lines be drawn perpendicular to the axes of X, Y and Z, respectively, right angled triangles will be formed, from which we have

$$x' = r' \cos X', \qquad y' = r' \cos Y', \qquad z' = r' \cos Z' \ldots\ldots(3).$$

In a similar way, we find

$$x'' = r'' \cos X'', \qquad y'' = r'' \cos Y'', \qquad z'' = r'' \cos Z''.$$

Substituting these values in equation (2), developing and arranging, we have

$$D^2 = (\cos^2 X' + \cos^2 Y' + \cos^2 Z')r'^2 + (\cos^2 X'' + \cos^2 Y'' + \cos^2 Z'')r''^2$$
$$- 2(\cos X' \cos X'' + \cos Y' \cos Y'' + \cos Z' \cos Z'')\, r'r'',$$

and substituting this in equation (1), we have

$$(\cos^2 X' + \cos^2 Y' + \cos^2 Z' - 1)r'^2 + (\cos^2 X'' + \cos^2 Y'' + \cos^2 Z'' - 1)r''^2$$
$$+ 2[\cos V - (\cos X' \cos X'' + \cos Y' \cos Y'' + \cos Z' \cos Z'')]r'r'' = 0.$$

Now since the points P and P′ were taken at pleasure, and since the angles V, X′, X″, &c., are entirely independent of the distances r' and r'', this equation will be true for any value of r' and r''; it is therefore an identical equation, in which the coefficients of r'^2, r''^2, &c., must be separately equal to 0; hence

$$\cos^2 X' + \cos^2 Y' + \cos^2 Z' = 1, \quad \cos^2 X'' + \cos^2 Y'' + \cos^2 Z'' = 1 \ldots(4),$$

$$\cos V = \cos X' \cos X'' + \cos Y' \cos Y'' + \cos Z' \cos Z'' \ldots\ldots(5).$$

From equations (4), we see that, *the sum of the squares of the cosines of the angles, which any right line makes with the co-ordinate axes, is equal to unity, or radius square.*

From equation (5), we see that, *the cosine of the angle formed by two right lines in space, is equal to the sum of the rectangles of the cosines of the angles formed by these lines with the co-ordinate axes.*

49. Since the point P is on the line AP, its co-ordinates x', y and z', must satisfy the equations of AP and give

$$x' = az', \qquad y' = bz' \ldots\ldots\ldots\ldots(1).$$

Substituting these values of x' and y' in the equation, Art. (42),

$$r'^2 = x'^2 + y'^2 + z'^2$$

and deducing the value of z', we have

$$z' = \frac{r'}{\sqrt{a^2 + b^2 + 1}}$$

and this value of z', in equations (1), gives

$$x' = \frac{ar'}{\sqrt{a^2 + b^2 + 1}}, \qquad y' = \frac{br'}{\sqrt{a^2 + b^2 + 1}}.$$

Substituting these values of x', y' and z', in equations (3), of the preceding article, we deduce

$$\cos X' = \frac{a}{\sqrt{a^2 + b^2 + 1}}, \qquad \cos Y' = \frac{b}{\sqrt{a^2 + b^2 + 1}}.$$

$$\cos Z' = \frac{1}{\sqrt{a^2 + b^2 + 1}}.$$

In a similar way, we may deduce

$$\cos X' = \frac{a'}{\sqrt{a'^2 + b'^2 + 1}}, \qquad \cos Y'' = \frac{b'}{\sqrt{a'^2 + b'^2 + 1}},$$

$$\cos Z'' = \frac{1}{\sqrt{a'^2 + b'^2 + 1}}.$$

Substituting these values in equation (5) of the preceding article, we have

$$\cos V = \pm \frac{aa' + bb' + 1}{\sqrt{a^2 + b^2 + 1}\sqrt{a'^2 + b'^2 + 1}} \ldots\ldots\ldots\ldots (3),$$

giving the double sign as the angle may be acute or obtuse.

If $V = 0$, $\cos V = 1$, hence

$$1 = \pm \frac{aa' + bb' + 1}{\sqrt{a^2 + b^2 + 1}\sqrt{a'^2 + b'^2 + 1}}.$$

Squaring both members, transposing and reducing, we obtain

$$(a - a')^2 + (b - b')^2 + (ab' - a'b)^2 = 0,$$

and since the first member is the sum of three positive terms, it can not be 0, unless each term is separately equal to 0; hence

$$a = a', \qquad b = b', \qquad ab' = a'b,$$

conditions deduced in article (47), the third evidently resulting from the other two.

If $V = 90°$, $\cos V = 0$; hence

$$aa' + bb' + 1 = 0,$$

which is *the equation of condition that two right lines, in space, shall be perpendicular to each other.* This equation being entirely different from, and independent of equation (5), Art. (47), shows that two lines may be perpendicular in space, without intersecting.

The angle, which the line AP makes with the plane XY, is evidently the complement of that which it makes with the axis of Z, and so with the other co-ordinate planes; hence if we denote these angles by U, U′ and U″, we have

$$\sin U = \cos Z', \qquad \sin U' = \cos Y', \qquad \sin U'' = \cos X',$$

or

$$\sin U = \frac{1}{\sqrt{a^2 + b^2 + 1}}, \qquad \sin U' = \frac{b}{\sqrt{a^2 + b^2 + 1}},$$

$$\sin U'' = \frac{a}{\sqrt{a^2 + b^2 + 1}};$$

expressions from which the angles, made by a given right line with the co-ordinate planes, may be determined.

50. Let

$$x = az + \alpha, \qquad y = bz + \beta,$$

be the general equations of a right line, in which a, b, α, and β, are undetermined, and let x', y', z' be the co-ordinates of a given point. If the line represented by the above equations passes through the given point, its co-ordinates must satisfy the equations and give the equations of condition

$$x' = az' + \alpha, \qquad y' = bz' + \beta.$$

If we subtract the last equations, member by member, from the first, we shall introduce the conditions thus expressed into the first, eliminate α and β, and obtain

$$x - x' = a\,(z - z') \ldots\ldots (1), \qquad y - y' = b\,(z - z') \ldots\ldots (2),$$

which are therefore, *the equations of a right line passing through a given point in space.*

In these equations a and b are still undetermined, as they should be, since an infinite number of lines may pass through the given point.

If the line is required to be parallel to a given line, the equations of which are

$$x = a'z + \alpha', \qquad y = b'z + \beta',$$

a and b will become known, since we must have, Art. (47),

$$a = a', \qquad b = b',$$

and by the substitution of these values, the line will be fully determined.

Find the equation of a right line, which shall pass through the point

$$x' = 2, \qquad y' = -3, \qquad z' = 1,$$

and be parallel to the line of which the equations are

$$x = 2z + 3, \qquad y = -z + 1.$$

51. If the line, represented by equations (1) and (2) of the preceding article, be subjected to the additional condition that it shall pass through the point whose co-ordinates are x'', y'' and z'', these co-ordinates must satisfy its equations and give the equations of condition

$$x'' - x' = a(z'' - z'), \qquad y'' - y' = b(z'' - z'),$$

from which we deduce

$$a = \frac{x'' - x'}{z'' - z'}, \qquad b = \frac{y'' - y'}{z'' - z'}.$$

Substituting these values in the equations (1) and (2), we have

$$x - x' = \frac{x'' - x'}{z'' - z'}(z - z'), \quad y - y' = \frac{y'' - y'}{z'' - z'}(z - z');$$

which are *the equations of a right line passing through two given points in space.*

Find the equations of a right line which shall pass through the two points

$$x' = 2, \quad y' = 0, \quad z' = 0; \qquad x'' = 0, \quad y'' = 3, \quad z'' = -1.$$

52. Curves, in space, may be represented in the same manner as the right line has been represented in Art. (44). Thus, if through a curve, cylinders be passed whose elements are perpendicular to the co-ordinate planes, these cylinders will be *the projecting cylinders of the curve,* and their intersections with the co-ordinate planes, *the projections of the curve,* either two of which being given, by their equations, the curve may be constructed by points, as in Art. (22).

53. The points of intersection of two curves, in space, may also be determined as in Art. (47), by combining their equations But as there will always be four equations, involving but three unknown quantities, proper values for the variables belonging to a common point, can not be found, unless an equation of condition, deduced as in that article, by eliminating x and y and equating the values of z, shall be satisfied.

To illustrate the intersection of two curves, let us take the equations

$$\left.\begin{aligned} 2z^2 - 3x &= 0 \ldots\ldots\ldots\ldots(1) \\ z^2 - 3y &= 0 \ldots\ldots\ldots\ldots(2) \end{aligned}\right\} \text{1st curve.}$$

$$\left.\begin{aligned} z^2 + 3x^2 - 12x + 9 &= 0 \ldots\ldots\ldots\ldots(3) \\ z^2 + 3y^2 - 6y &= 0 \ldots\ldots\ldots\ldots\ldots\ldots(4) \end{aligned}\right\} \text{2nd curve.}$$

If we combine equations (1) and (3), and deduce the values of x and z, we have

$x = 2, \qquad z = \pm \sqrt{3}$

$x = \frac{3}{2}, \qquad z = \pm \frac{3}{2}.$

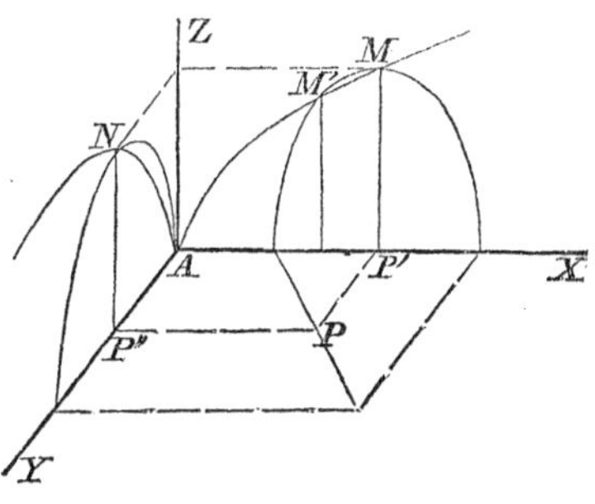

These values of x and z are evidently the co-ordinates of the points M and M', in which the projections of the curves on the plane XZ intersect.

Combining equations (2)and (4), we obtain

$$y = 1, \qquad z = \pm \sqrt{3};$$

$$y = 0, \qquad z = \pm 0,$$

and these are the co-ordinates of the points, A and N, common to the projections of the curves on the plane YZ. The second values of z, in the two cases, being unequal, can not, with the corresponding values of x and y, satisfy all four equations at the same time and therefore do not belong to a point common to the two curves. The first values of z, viz. $z = \pm \sqrt{3}$, are the same in both cases and therefore taken with $x = 2$, and $y = 1$, are the co-ordinates of two points in which the curves intersect, one of these points being above, and the other the same distance below the plane XY, at P.

The same result may be otherwise obtained thus: Combine equations (1) and (3) and eliminate x, thus deducing an equation involving z. Combine equations (2) and (4) and eliminate y, thus deducing another equation in z; and since there can be no common point unless these equations give equal values for z, it follows (the second member of both being 0), that for each equal value of z the first members will have a common divisor of the form $z - a$; hence, if we seek the greatest common divisor of these first members and place it equal to 0, the roots of the resulting equation will give all the values of z which will satisfy both equations. Those which give real values of x in (1) and (3), and real values of y in (2) and (4), will correspond to points of intersection. By applying this process to the above equations we find for the greatest common divisor $z^2 - 3$, which placed equal to 0, gives

$$z = \pm \sqrt{3},$$

the same values before found.

If only the form of the equations of two curves should be given, the constants which enter them being arbitrary, x and y may be eliminated, as above, and then such values may often be assigned to these constants, as to give the first members of the resulting equations in z, a common divisor of the first or higher degree, thus causing the two curves to intersect in one or more points.

OF THE PLANE.

54. *The equation of a surface is an equation which expresses the relation between the co-ordinates of every point of the surface.*

A plane surface may be generated, by moving a straight line, so as to touch another straight line, and have all of its positions parallel to its first position. The moving line is called *the generatrix*; and the line on which it moves, or which directs its motion, *the directrix.*

55. Let

$$y = a'x + b' \ldots\ldots\ldots\ldots (1),$$

be the equation of any right line, DB, in the plane XY, and let

$$x = az + \alpha, \qquad y = bz + \beta \ldots\ldots\ldots\ldots (2),$$

be the equations of a right line in space, which is to be moved on the line DB, so as to generate a plane. Since the moving line must always be parallel to its first position, a and b will remain the same in all of its positions, while α and β will change, as the line is moved from one position to another. But α and β are the co-ordinates of the point, in which the line pierces the co-ordinate plane XY, Art. (46), and since this point must be on the line DB, the values of α and β, deduced from equations (2), must, in all positions of the generatrix, satisfy equation (1), when substituted for the variables. The values, thus deduced, are

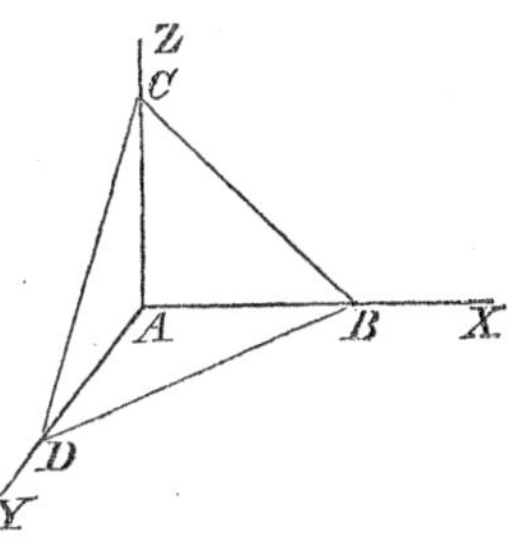

$$\alpha = x - az, \qquad \beta = y - bz,$$

and these, substituted for x and y in equation (1), give

$$y - bz = a'(x - az) + b' \ldots\ldots\ldots\ldots (3),$$

which expresses a relation between the co-ordinates of the different points of the generatrix, in all of its positions; it is, therefore, *the equation of a plane.* If this equation be solved with reference to z, and the coefficients of x and y be placed equal to c and d, respectively, and the absolute term equal to g, we have

$$z = cx + dy + g \ldots\ldots\ldots\ldots (4),$$

a form analogous to that of the right line, Art. (24).

Since this equation contains three variables, either two may be assumed at pleasure and the corresponding value of the third deduced; the three, taken together, will be the co-ordinates of a point of the plane, which may be constructed as in Art. (40), and as any number of its points may be determined in the same way, the plane will evidently be given when the constants which enter its equation are known.

And, in general, any surface will be given, analytically, when the form of its equation and the constants which enter it are known.

56. The intersection of a plane with either co-ordinate plane is called a *trace of the plane.*

For every point of the plane, which lies in the co-ordinate plane XZ, y must be equal to 0. Substituting this value for y, in equation (4) of the preceding article, we obtain

$$z = cx + g \dots\dots\dots\dots (1),$$

in which x and y can only belong to points of the plane lying in the plane XZ. This is then the equation of the trace, BC, on the plane XZ.

In the same way, for all points of the plane, in YZ, x must be equal to 0; whence

$$z = dy + g \dots\dots\dots\dots (2),$$

is the equation of the trace, DC, on the plane YZ.

By making $z = 0$, we obtain

$$cx + dy + g = 0,$$

for the equation of the trace, BD, on the plane XY.

For all points in the axis of Z, x and y must be equal to 0.

Substituting these values for x and y in equation (4), we find

$$z = g,$$

which is the distance AC, cut off by the plane on the axis of Z.

In a similar way, we find the distances cut off on the axes of X and Y

$$x = -\frac{g}{c} = \text{AB}, \qquad y = -\frac{g}{d} = \text{AD}.$$

If $g = 0$, these distances become 0, the plane will pass through the origin, and its equation become

$$z = cx + dy,$$

without an absolute term, as it should be, since the co-ordinates of the origin will then satisfy the equation.

If $c = 0$, the distance AB becomes infinity, and the plane is parallel to the axis of X, or perpendicular to the co-ordinate plane YZ, and its equation becomes

$$z = dy + g,$$

the same as that of the trace on ZY. It should be remarked, however, that for the plane, x may have any value, or *is indeterminate*, since its coefficient c is 0; while for the trace, x must be equal to 0, as we have seen.

If $d = 0$, the distance AD becomes infinity, and the equation of the plane perpendicular to XZ,

$$z = cx + g, \qquad y \text{ indeterminate.}$$

In the same way, if equation (3), Art. (55), had been solved with reference to y or x, it might be shown that the equation of a plane perpendicular to XY, would be *the same as that of its trace, z being indeterminate.*

57. Every equation of the first degree between three variables, will be a particular case of the general equation

$$Ax + By + Cz + D = 0,$$

and this, when solved with reference to z, gives

$$z = -\frac{A}{C}x - \frac{B}{C}y - \frac{D}{C},$$

an equation of the same nature and form as

$$z = cx + dy + g \ldots\ldots\ldots\ldots (1),$$

and will therefore represent a magnitude of the same kind; that is, *every equation of the first degree between three variables is the equation of a plane*, and when solved with reference to z, will appear under the form (1).

58. Let

$$x = az + \alpha, \qquad y = bz + \beta \ldots\ldots\ldots\ldots (1),$$

be the equations of a right line, and

$$z = cx + dy + g \ldots\ldots\ldots\ldots (2),$$

the equation of a plane. Those values of x, y and z which, when taken together, will satisfy these three equations at the same time, must be the co-ordinates of a point common to the line and plane. Therefore, by combining the equations and deducing the values of x, y and z, we shall obtain the co-ordinates of the point in which the line pierces the plane. Substituting the values of x and y, from equations (1), in equation (2), we find

$$z = \frac{\alpha c + \beta d + g}{1 - ac - bd},$$

and by the substitution of this value of z in equations (1), we may deduce the corresponding values of x and y. If

$$1 - ac - bd = 0,$$

the values of z, x and y will become infinite, the point in which the line pierces the plane will be at an infinite distance, and *the line will be parallel to the plane*. The last equation is then the analytical condition that a right line shall be parallel to a plane; or, that a right line, having one point in a plane, shall lie wholly in the plane.

In the same way, the points in which any line, in space, pierces a surface may be found; since the two equations of the line, with the equation of the surface, will always give three equations, by the combination of which, values of the three variables may be deduced which will satisfy the equations at the same time. The number of sets of real values thus found will indicate the number of common points.

59. Let

$$z = cx + dy + g,$$

be the equation of a plane, and suppose any straight line to be drawn perpendicular to the plane. If through the point where the plane cuts the axis of Z, a line be drawn parallel to the given line, its equations will be of the form

$$x = az + \alpha, \qquad y = bz + \beta,$$

in which a and b are the same as in the equations of the given line, Art. (47). Since this second line is also perpendicular to the plane, it must be perpendicular to the traces, BC and DC, which are two lines of the plane passing through its foot. The equations of the trace BC, Art. (56), may be put under the form

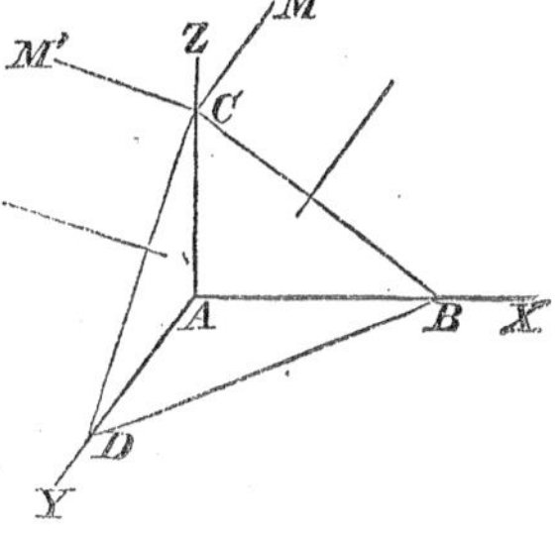

$$x = \frac{1}{c}z - \frac{g}{c}, \qquad y = 0.z,$$

since the projection of BC, on the plane YZ, coincides with the axis of Z.

The general equation of condition that the right line shall be perpendicular to the trace is, Art. (49),

$$1 + aa' + bb' = 0 \text{............} (1),$$

in which, from the above equations of the trace, we must have

$$a' = \frac{1}{c} \qquad b' = 0.$$

Substituting these values in equation (1), we obtain

$$1 + \frac{a}{c} = 0 \text{............} (2), \qquad \text{or} \qquad a = -c$$

for the condition that the line shall be perpendicular to the trace.

In a similar way, for the trace DC, we have

$$b' = \frac{1}{d} \qquad a' = 0,$$

and these, in equation (1), give

$$1 + \frac{b}{d} = 0 \text{............} (3), \qquad \text{or} \qquad b = -d.$$

$$a = -c \qquad b = -d$$

are then the analytical conditions that a straight line shall be perpendicular to a plane.

Condition (2) proves also that the projection CM is perpendicular to the trace BC, Art. (28); and condition (3) proves that the projection CM′ is perpendicular to DC. Hence, *if a right line is perpendicular to a plane, its projections are perpendicular to the traces of the plane, respectively.*

60. Let x', y', z', be the co-ordinates of a given point, and

$$z = cx + dy + g \text{............} (1),$$

the equation of a given plane. The equations of a right line passing through the given point will be, Art. (50),

$$x - x' = a(z - z') \qquad y - y' = b(z - z') \text{............} (2).$$

If this line is required to be perpendicular to the plane, we must have, by the preceding article,

$$a = -c, \qquad b = -d.$$

Substituting these values in equations (2), we have

$$x - x' = -c(z - z'), \qquad y - y' = -d(z - z') \dots\dots(3),$$

for the equations of a right line passing through a given point and perpendicular to the plane.

The point, in which this perpendicular pierces the plane, may be found as in Art. (58), by combining equations (3) with equation (1); and the distance between this and the given point, or the length of the perpendicular, by means of the formula of Art. (42).

Find the equations of a straight line passing through a point whose co-ordinates are

$$x' = -2, \qquad y' = 1, \qquad z' = 3,$$

and perpendicular to the plane whose equation is

$$2x - 3y + 4z + 1 = 0.$$

Find also the point in which the line pierces the plane, and the length of the perpendicular.

61. The angle, made by a straight line with a plane, is the same as the angle included between the line and its projection on the plane. Therefore, if through any point of the line a perpendicular be drawn to the plane, this perpendicular, a portion of the line and its projection on the plane, will form a right angled triangle, of which the angle at the base will be the angle made by the line and plane, and the angle at the vertex, its complement.

Denote the first angle by A, and the angle formed by the given line and the perpendicular by V. Then, the line being represented by equations (1) and (2), Art. (44), and the plane by equation (4), Art. (55), the perpendicular will be represented by equations (3) of the preceding article, and by substituting $-c$ for a', and $-d$ for b' in the formula (3), of Art. (49,) we have

$$\cos V = \pm \frac{1 - ac - bd}{\sqrt{1 + a^2 + b^2}\sqrt{1 + c^2 + d^2}} = \sin A,$$

from which we determine the sine of A, and thence the angle itself.

If

$$1 - ac - bd = 0,$$

the angle becomes 0, and the line is parallel to the plane, a condition before determined, Art. (58).

62. Let

$$z = cx + dy + g \ldots\ldots\ldots\ldots (1),$$

$$z = c'x + d'y + g' \ldots\ldots\ldots\ldots (2),$$

be the equations of two planes. Those values of x, y and z which will satisfy both of these equations, at the same time, must belong to points common to the two planes. If then we combine these equations, x, y and z in the result can only belong to the line of intersection; and if one of the variables, as z, be eliminated, we have

$$(c - c')x + (d - d')y + g - g' = 0 \ldots\ldots\ldots\ldots (3),$$

which must be the equation of the projection of this line of intersection on the plane XY. In the same way, if the equations be combined and x be eliminated, the result will be the equation of the projection of the line of intersection on the plane YZ. Two projections being thus determined, the line will be known.

If such a relation exists between c, c', d and d', that no values of x and y will satisfy equation (3), the planes can not intersect, but must be parallel. This can only be the case when $c = c'$ and $d = d'$, as we shall then have

$$g - g' = 0,$$

which can not be if the planes are different; hence,

$$c = c', \qquad d = d',$$

are *the analytical conditions that two planes shall be parallel.*

By referring to the equations of the traces of these planes, we see that $c = c'$ is the condition that the traces on the plane ZX shall be parallel, Art. (28), and that $d = d'$ is the condition that the traces on the plane ZY shall be parallel; hence, *if two planes are parallel, their traces are parallel.*

If the plane represented by equation (1) is parallel to the co-ordinate plane XY, its traces on XZ and YZ must be parallel, respectively, to the axes of X and Y; hence, by a reference to the equations of these traces, Art. (56), we see that

$$c = 0, \qquad d = 0,$$

and that equation (1) reduces to

$$z = g, \qquad x \textit{ and } y \textit{ indeterminate},$$

for the equation of a plane parallel to the co-ordinate plane XY.

If $g = 0$, also, we have

$$z = 0, \qquad x \textit{ and } y \textit{ indeterminate},$$

for the equation of the co-ordinate plane XY.

If the plane represented by (1) is parallel to the co-ordinate plane YZ, its traces on XZ and XY must be parallel to the axes of Z and Y which requires

$$\frac{1}{c} = 0, \qquad -\frac{d}{c} = 0.$$

These values, substituted in equation (1), placed under the form

$$x = \frac{1}{c}z - \frac{d}{c}y - \frac{g}{c},$$

give

$$x = -\frac{g}{c} \quad \text{or} \quad x = h, \quad y \textit{ and } z \textit{ indeterminate},$$

for the equation of a plane parallel to YZ, and at a distance from it equal to $-\frac{g}{c} = h$.

If $g = 0$, also, we have

$$x = 0 \qquad y \textit{ and } z \textit{ indeterminate},$$

for the equation of the plane YZ; and similar equations may be found for a plane parallel to XZ, and for the plane XZ itself.

The preceding method of finding the intersection of two planes is applicable to any surfaces whatever. Thus: Combine the equations of the surfaces, and eliminate one of the variables, the result will be the equation of the projection of the intersection on the plane of the other two variables. Combine the equations again and eliminate another variable, the result will be the equation of the projection on another plane, and the intersection will be thus determined.

Find the intersection of the two planes whose equations are

$$2x - 3y + 2z = 0,$$

$$x + 2y - 3z + 1 = 0.$$

63. If through any point, within the angle included by two planes, a line be drawn perpendicular to each plane, the angle included by one of these lines and the prolongation of the other, will be equal to the angle included by the planes. Let the equa-

tions of the planes be the same as in the preceding article, then the equations of the perpendiculars will be, Art. (60),

$$x - x' = - c\,(z - z'), \qquad y - y' = - d\,(z - z'),$$

$$x - x' = - c'\,(z - z'), \qquad y - y' = - d'\,(z - z'),$$

If we denote the angle which these lines make, by A, and then substitute $-c$ and $-c'$ for a and a', and $-d$ and $-d'$ for b and b', in formula (3), Art. (49), we have

$$\cos A = \pm \frac{1 + cc' + dd'}{\sqrt{1 + c^2 + d^2}\,\sqrt{1 + c'^2 + d'^2}} \text{ ....(1)},$$

from which we deduce the value of cos A, and thence of A itself, which will express the number of degrees, &c., contained in the angle of the planes.

If the two planes are parallel, we have $A = 0$, $\cos A = 1$. By substituting this value of cos A, clearing of denominators, &c., as in Art. (49), we may deduce the same equations of condition as in the preceding article.

If the two planes are perpendicular to each other, we must have $A = 90°$, $\cos A = 0$, which requires

$$1 + cc' + dd' = 0,$$

the equation of condition that two planes shall be perpendicular to each other.

If the first plane coincides with the plane XY, we have, from the preceding article,

$$c = 0 \qquad d = 0,$$

and cos A reduces to

$$\cos X'' = \frac{1}{\sqrt{1 + c'^2 + d'^2}},$$

for the cosine of the angle made by the second plane with the co-ordinate plane XY.

If the same plane coincides with the plane YZ, we have

$$\frac{1}{c} = 0, \qquad \frac{d}{c} = 0,$$

and these values substituted in equation (1), first placing it under the form,

$$\cos A = \pm \frac{\frac{1}{c} + c' + \frac{d}{c} d'}{\sqrt{\frac{1}{c^2} + 1 + \frac{d^2}{c^2}}\sqrt{1 + c'^2 + d'^2}}$$

reduce it to

$$\cos Y'' = \frac{c'}{\sqrt{1 + c'^2 + d'^2}},$$

for the cosine of the angle made by the second plane with the plane YZ.

If the plane coincides with XZ, we have

$$\frac{1}{d} = 0, \qquad \frac{c}{d} = 0,$$

and equation (1) may be reduced to

$$\cos Z'' = \frac{d'}{\sqrt{1 + c'^2 + d'^2}}.$$

In the same way, if the second plane be made to coincide, in succession, with each co-ordinate plane, we may deduce for the angles X', Y' and Z', made by the first plane with the co-ordinate planes

$$\cos X' = \frac{1}{\sqrt{1 + c^2 + d^2}}, \qquad \cos Y' = \frac{c}{\sqrt{1 + c^2 + d^2}},$$

$$\cos Z' = \frac{d}{\sqrt{1 + c^2 + d^2}}.$$

If both members of these three equations be squared and the results added, member to member, we find

$$\cos^2 X' + \cos^2 Y' + \cos^2 Z' = 1.$$

If the values of cos X′ and cos X″ be multiplied together, also cos Y′ and cos Y″, cos Z′ and cos Z″ and the three products added, we obtain

$$\cos X' \cos X'' + \cos Y' \cos Y'' + \cos Z' \cos Z'' = \cos A,$$

an expression for the cosine of the angle formed by two planes, in terms of the cosines of the angles made by the planes with the co-ordinate planes.

64. Let x', y', z', be the co-ordinates of a given point, and

$$z = cx + dy + g \ldots\ldots\ldots\ldots (1),$$

the general equation of a plane, in which c, d and g are arbitrary constants. If the given point is in this plane its co-ordinates must satisfy the equation and give the equation of condition,

$$z' = cx' + dy' + g.$$

Subtracting this equation, member by member, from (1), we introduce the condition into that equation and obtain,

$$z - z' = c(x - x') + d(y - y'),$$

for the equation of a plane passing through a given point, in which c and d are still arbitrary.

65. If the plane, represented by equation (1) of the preceding

article, be required to contain the three given points x', y', z', x'', y'', z'', and x''', y''', z''', these co-ordinates, when substituted in succession for the variables, must satisfy the equation and give the three equations of condition,

$$z' = cx' + dy' + g,$$
$$z'' = cx'' + dy'' + g,$$
$$z''' = cx''' + dy''' + g.$$

From these three equations, the values of the three constants c, d and g may be determined, and substituted in equation (1). The result will be the equation of a plane passing through three given points.

Find the equation of a plane passing through the three points,

$$x' = 1, \quad y' = 0, \quad z' = -3;$$
$$x'' = 2, \quad y'' = 1, \quad z'' = 1;$$
$$x''' = 0, \quad y''' = 2, \quad z''' = 0.$$

TRANSFORMATION OF CO-ORDINATES.

66. In developing and discussing the properties of lines and surfaces, it is often of great advantage to change the reference of their points from one system of co-ordinates axes or planes to another. The system from which the change is made is called *the primitive system;* the one to which it is made is *the new system;* and changing the reference of points, from one system of co-ordinate axes or planes to another, is called *the transformation of co-ordinates.*

If a line or surface be given by its equation, and it be required to change the reference of its points to a new system of co-ordinate axes or planes; it is only necessary to deduce values for *the primitive co-ordinates* in terms of *the new*, and to substitute these values for the variables in the given equation. The result, expressing a relation between the new co-ordinates of the points, will of course be the equation of the line referred to the new system.

From the nature of this operation, it is evident that no change whatever takes place, either in the nature or extent of the line or surface.

67. Let AX and AY be any set of co-ordinate axes, and AX′ and AY′ any other set having the same origin. Denote the angle included between AX and AY by β, and let α and α' denote the angles made by AX′ and AY′, respectively, with AX. Let AP $= x$ and MP $= y$ be the co-ordinates of any point, as M, when referred to the first set, and let AP′ $= x'$ and MP′ $= y'$ be the co-ordinates of the same point referred to the second set. Through P′ draw P′R parallel to AX and P′S parallel to AY.

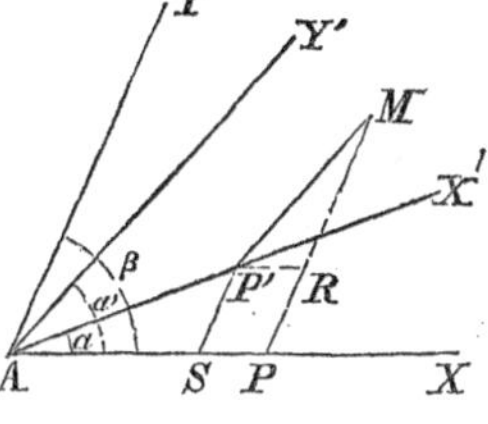

In the triangle ASP′, the angle

$$AP'S = P'AY = \beta - \alpha, \qquad \sin ASP' = \sin YAX = \sin \beta,$$

and since the sides are as the sines of their opposite angles, we have the two proportions,

$$AS : AP' :: \sin(\beta - \alpha) : \sin ASP' \text{ or } \sin \beta,$$

$$P'S : AP' :: \sin \alpha : \sin \beta;$$

whence

$$AS = \frac{x' \sin(\beta - \alpha)}{\sin \beta}, \qquad P'S = \frac{x \sin \alpha}{\sin \beta}.$$

In the triangle P′RM,

$$P'MR = YAY' = \beta - \alpha', \qquad MP'R = Y'AX = \alpha',$$
$$MRP' = P'SA,$$

and we have the proportions

$$P'R : P'M :: \sin(\beta - \alpha') : \sin \beta,$$
$$MR : P'M :: \sin \alpha' : \sin \beta;$$

whence

$$P'R = \frac{y' \sin(\beta - \alpha')}{\sin \beta}, \qquad MR = \frac{y' \sin \alpha'}{\sin \beta}.$$

We have also

$$AP = AS + P'R, \qquad MP = P'S + MR.$$

Substituting, in these equations, the values above deduced, we have

$$x = \frac{x' \sin(\beta - \alpha) + y' \sin(\beta - \alpha')}{\sin \beta},$$

$$y = \frac{x' \sin \alpha + y' \sin \alpha'}{\sin \beta},$$

in which the values of the primitive co-ordinates are expressed in terms of the new and constants; and these are the formulas, for passing from any system of rectilineal co-ordinates to another having the same origin.

If the new origin is different from the primitive, at A′, for instance, it is evident that we have simply to add to the above values, a' and b', the co-ordinates of the new origin referred to the primitive system. We thus obtain

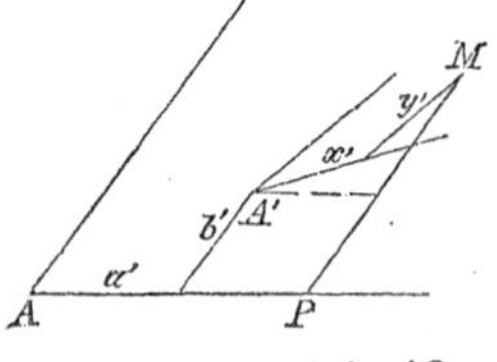

$$x = a' + \frac{x' \sin(\beta - \alpha) + y' \sin(\beta - \alpha')}{\sin \beta},$$

$$y = b' + \frac{x' \sin \alpha + y' \sin \alpha'}{\sin \beta},$$

general formulas for passing from one system of rectilineal co-ordinates to any other, in the same plane.

If the new axes of co-ordinates are required to be parallel to the primitive, we have

$$\alpha = 0, \qquad \alpha' = \beta, \qquad \sin \alpha = 0, \qquad \sin \alpha' = \sin \beta,$$

and the above formulas reduce to

$$x = a' + x', \qquad y = b' + y' \ldots\ldots\ldots\ldots (2),$$

formulas for passing from any set of co-ordinate axes to a parallel set, in the same plane.

If the primitive axes are perpendicular to each other, we have

$$\beta = 90°, \quad \sin \beta = 1, \quad \sin (\beta - \alpha) = \cos \alpha,$$
$$\sin (\beta - \alpha') = \cos \alpha',$$

and formulas (1), reduce to

$$\left.\begin{array}{l} x = a' + x' \cos \alpha + y' \cos \alpha' \\ y = b' + x' \sin \alpha + y' \sin \alpha' \end{array}\right\} \ldots\ldots\ldots\ldots (3),$$

formulas for passing from a system of rectangular co-ordinate axes to an oblique system, in the same plane.

If the primitive axes are perpendicular to each other, and also the new, we have

$$\beta = 90°, \quad \sin \beta = 1, \quad \alpha' = 90° + \alpha, \quad \sin \alpha' = \cos \alpha,$$

$$\sin (\beta - \alpha) = \cos \alpha, \qquad \sin (\beta - \alpha') = \sin(-\alpha) = -\sin \alpha,$$

and formulas (1) reduce to

$$\left.\begin{array}{l} x = a' + x' \cos \alpha - y' \sin \alpha \\ y = b' + x' \sin \alpha + y' \cos \alpha \end{array}\right\} \ldots\ldots\ldots\ldots (4),$$

formulas *for passing from a system of rectangular co-ordinate axes to another system, also rectangular, in the same plane.*

If the new axes, only, are perpendicular to each other, we have

$$\alpha' = 90° + \alpha, \qquad \sin\alpha' = \cos\alpha, \qquad \cos\alpha' = -\sin\alpha,$$

and formulas (1) reduce to

$$\left.\begin{aligned} x &= a' + \frac{x'\sin(\beta-\alpha) - y'\cos(\beta-\alpha)}{\sin\beta} \\ y &= b' + \frac{x'\sin\alpha + y'\cos\alpha}{\sin\beta} \end{aligned}\right\}\ldots\ldots(5),$$

formulas *for passing from a system of oblique co-ordinate axes to a rectangular system, in the same plane.*

If the new origin be the same as the primitive, a' and b' in each of the above formulas will be equal to 0.

68. We may illustrate the use of the formulas of the preceding article by the following

Examples.

1. Let

$$x^2 + y^2 = R^2 \ldots\ldots\ldots\ldots(1),$$

be the equation of a circle referred to its centre and rectangular co-ordinate axes, Art. (35), and let it be proposed to change the reference to a parallel set having the origin at the point C.

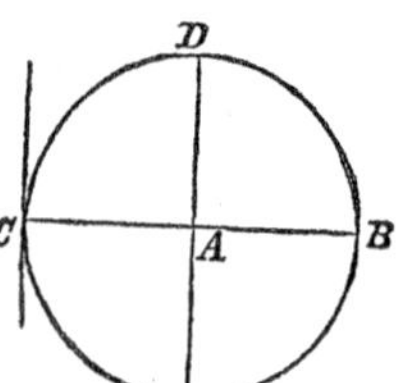

The co-ordinates of the new origin will be

$$a' = -R, \qquad b' = 0,$$

and these values, in formulas (2), reduce them to

$$x = -R + x', \qquad y = y'.$$

Substituting these last values for x and y in equation (1), and reducing, we obtain

$$y'^2 = 2Rx' - x'^2,$$

an equation before found in Art. (38).

2. Let

$$y = ax + b \ldots\ldots\ldots\ldots(2),$$

be the equation of the right line A′B, referred to the rectangular axes AX and AY, and let it be proposed to find the equation of the same line referred to the axes A′X′ and A′Y′, also at right angles, the axis of X′ making an angle of 45° with the axis of X and having the new origin at A′, the point where the given line cuts the axis of Y. The general formulas to be used in this case are formulas (4), in which

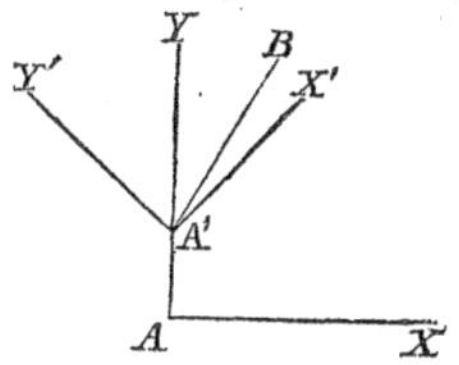

$$a' = 0, \qquad b' = b, \qquad \sin \alpha = \cos \alpha.$$

These values reduce the formulas to

$$x = (x' - y') \cos \alpha, \qquad y = b + (x' + y') \cos \alpha,$$

and substituting these values for x and y, in equation (2), we have

$$b + (x' + y') \cos \alpha = a(x' - y') \cos \alpha + b,$$

or reducing,

$$y' = \frac{a - 1}{a + 1} x'.$$

69. Let AX and AY be a set of rectangular co-ordinate axes,

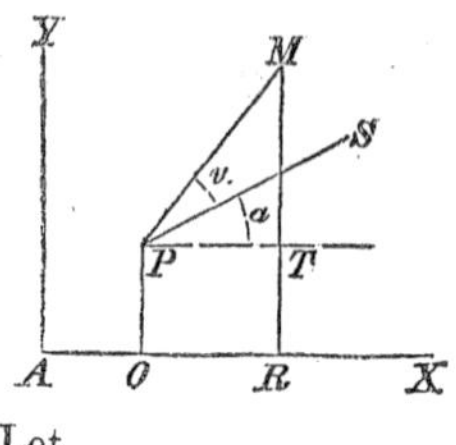

and M any point referred to them by the co-ordinates $AR = x$, and $MR = y$; and let P be the pole, and PS the fixed line, to which the point is referred by the radius vector $PM = r$ and the angle $MPS = v$, Art. (18). Let

$$AO = a', \qquad OP = b', \qquad SPT = \alpha.$$

In the right angled triangle, MTP, we have

$$PT = r \cos (v + \alpha), \qquad MT = r \sin (v + \alpha).$$

Substituting the above values in the equations

$$AR = AO + PT, \qquad MR = OP + MT,$$

we have

$$x = a' + r \cos (v + \alpha), \qquad y = b' + r \sin (v + \alpha) \ldots\ldots (1),$$

which are general formulas, *for passing from a system of rectangular co-ordinates to a system of polar co-ordinates, in the same plane.*

The fixed line is generally taken parallel to the axis of X, in which case $\alpha = 0$, and formulas (1) reduce to

$$x = a' + r \cos v, \qquad y = b' + r \sin v \ldots\ldots\ldots\ldots (2).$$

If the pole is at the origin, we have $a' = 0$, $b' = 0$.

From the second of equations (1), we deduce

$$r = \frac{y - b'}{\sin (v + \alpha)},$$

in which, if $y > b'$, $y - b'$ is positive, the point M is above the line PT, and $\sin (v + \alpha)$ also positive; hence, the value of r will be essentially positive.

If $y < b'$, $y - b'$ is negative, the point M is below PT, $\sin(v + \alpha)$ also negative, and the value of r positive.

The value of the radius vector is therefore always positive. Hence, if in discussing the equation of a line referred to a fixed point and fixed right line, usually called *the polar equation of the line*, a negative value of the radius vector is found, it must be rejected, as there can be no corresponding point.

70. To illustrate the principles of the preceding article, let it be proposed to determine and discuss the polar equation of the circle. Its equation referred to the rectangular axes AX and AY, is

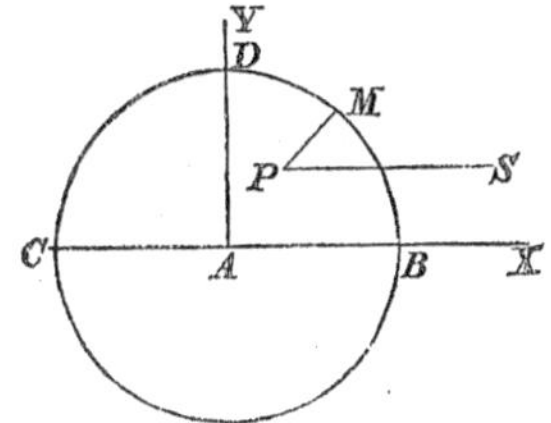

$$x^2 + y^2 = R^2 \ldots\ldots\ldots\ldots (1).$$

Suppose the fixed line PS, from which the angle v is estimated, is parallel to the axis of X, we must then use formulas (2). Squaring the values of x and y, we have

$$x^2 = a'^2 + 2a'r \cos v + r^2 \cos^2 v,$$

$$y^2 = b'^2 + 2b'r \sin v + r^2 \sin^2 v.$$

Substituting these values in equation (1), recollecting that

$$\sin^2 v + \cos^2 v = 1,$$

and reducing, we obtain

$$r^2 + 2(a' \cos v + b' \sin v)r + a'^2 + b'^2 - R^2 = 0 \ldots\ldots\ldots\ldots (2),$$

for the general polar equation of the circle.

By attributing particular values to a' and b', the pole may be placed at any point in the plane of the circle.

If the pole be placed at C, we must have

$$a' = -R, \qquad b' = 0,$$

and these, in equation (2), give

$$r^2 - 2R \cos vr = 0.$$

This equation gives two values of r,

$$r = 0, \qquad r = 2R \cos v.$$

These two values of r represent the distances from the pole to the points in which the radius vector, making any angle v, cuts the circle. Since the pole is on the curve, one of these values is necessarily 0, whatever be the angle v. The second may then represent any radius vector as CM.

If in this second value $v = 0$, we have $\cos v = 1$, and

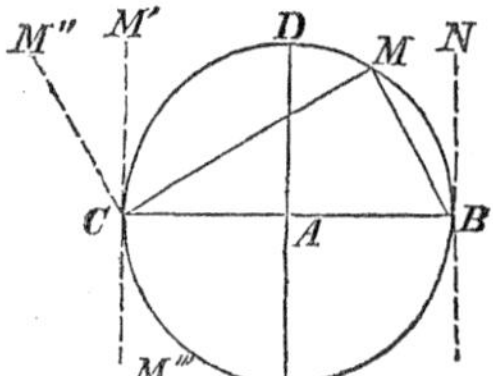

$$r = 2R = CB,$$

which gives the point B. As v increases, $\cos v$ will remain positive until $v = 90°$, in which case $\cos v = 0$, r becomes 0, and the radius vector takes the position CM′ tangent to the circle at C. As v increases beyond 90°, its cosine becomes negative, the value of r is negative and gives no point of the curve, until v becomes equal to 270°, when $\cos v = 0$ and $r = 0$, taking the position CM‴. As v increases beyond 270°, its cosine is positive, r is positive and gives points of the curve until $v = 360°$, when we again have $r = CB$.

From this we see that as v increases from 0 to 90°, we obtain all the points in the semi-circumference BDC, that no points of the curve are on the left of the line M′M‴, and that as v increases from 270° to 360°, we obtain all the points in the other semi-circumference.

The second value of r is readily verified, since in the right angled triangle CMB, we have

$$CM = CB \cos BCM \qquad \text{or} \qquad r = 2R \cos v.$$

If the pole is placed at B, we have

$$a' = R, \qquad b' = 0,$$

and equation (2) gives the two values

$$r = 0, \qquad r = -2R \cos v.$$

The second value of r will be negative for all values of v less than 90° or greater than 270°, and positive for all values from 90° to 270°.

If the pole is placed at the centre, we have

$$a' = 0, \qquad b' = 0,$$

and equation (2) reduces to

$$r = R,$$

v being indeterminate, since its coefficient is equal to 0.

71. By reflecting upon the discussion contained in the three preceding articles, we see that two classes of propositions may arise in the transformation of co-ordinates.

First; when it is proposed to change the reference from a given set of co-ordinate axes to another set, the exact position of which is known. In this case the constants which enter the values of the primitive co-ordinates are given.

Second; when it proposed to change from a given set to another, the position of which is to be determined, so that the resulting equation shall assume a certain form, or the new set fulfil certain conditions. In this case, the constants above referred to are arbitrary, and by assigning values to them, as many reasonable conditions may be introduced as there are such constants, and the position of the new co-ordinate axes thus determined.

72. Let AX, AY and AZ, be three co-ordinate axes at right angles to each other, and AX′, AY′ and AZ′, three oblique axes having the same origin. Denote the angles made by the new axis AX′ with the three primitive axes of X, Y and Z, respectively, by X, Y and Z, those made by the axis AY′ with the same, by X′, Y′ and Z′, and those made by AZ′, by X″, Y″ and Z″.

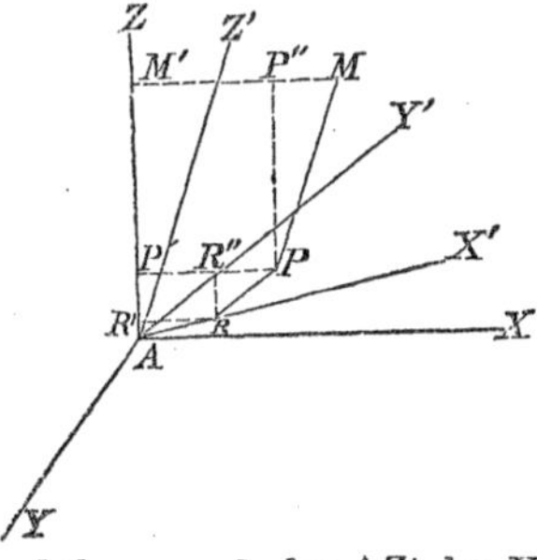

Let M be any point, in space, referred to the primitive planes by the co-ordinates x, y and z. Through this point draw the line MP parallel to AZ′, until it pierces the new plane X′Y′, in the point P; through this last point, draw PR parallel to AY′, until it intersects the new axis of X′, in R; then

$$AR = x', \qquad PR = y', \qquad MP = z',$$

are the co-ordinates of the point M referred to the oblique co-ordinate planes. Through the points M, P and R, pass planes parallel to the plane XY, intersecting the axis of Z in M′, P′ and R′. AM′ is equal to z, and the lines AR, RP and PM, are the hypothenuses of right angled triangles, the bases of which are AR′, RR″ and PP″, and the angles at the bases, Z, Z′ and Z″. From these triangles we have

$$AR' = AR \cos Z, \quad RR'' = RP \cos Z', \quad PP'' = MP \cos Z''.$$

Substituting these values for their equals in the equation

$$AM' = AR' + R'P' + P'M',$$

and for AM′, AR, RP and MP, their values, we have,

$$z = x' \cos Z + y' \cos Z' + z' \cos Z''.$$

In a similar way, by drawing lines through the point M respectively parallel to the new axes of X′ and Y′, we may deduce

$$x = x' \cos X + y' \cos X' + z' \cos X'',$$

$$y = x' \cos Y + y' \cos Y' + z' \cos Y''.$$

These three equations taken together express the values of the primitive co-ordinates in terms of the new, and are the formulas for changing the reference of points from a set of co-ordinate planes at right angles, to another set oblique to each other, having the same origin.

If the origin be also changed to a point whose co-ordinates are a, b and c, these formulas become

$$x = a + x' \cos X + y' \cos X' + z' \cos X'',$$

$$y = b + x' \cos Y + y' \cos Y' + z' \cos Y'', \ldots\ldots\ldots\ldots(1).$$

$$z = c + x' \cos Z + y' \cos Z' + z' \cos Z'',$$

In these formulas there are twelve constants; but since the angles X, Y, Z, &c., made by each of the new axes with the primitive, must fulfil the condition expressed in equation (4), Art. (48), thus forming three equations of condition, we can, by means of these constants, introduce only nine independent conditions.

If the new axes are also perpendicular to each other, we shall have the cosines of the angles, included between each set of two, equal to 0. Placing the expressions for these cosines, Art. (48), each equal to 0, we have three more equations of condition existing between the arbitrary constants.

If the new axes are parallel to the primitive, we have

$$X = 0, \qquad Y' = 0, \qquad Z'' = 0,$$

and each of the other angles equal to 90°, hence the above formulas reduce to

$$x = a + x', \qquad y = b + y', \qquad z = c + z' \ldots\ldots(2),$$

which are *the formulas for passing from a set of planes at righi angles, to a parallel set.*

73. Let M be any point, in space, referred to the three rectangular co-ordinate planes, by the co-ordinates

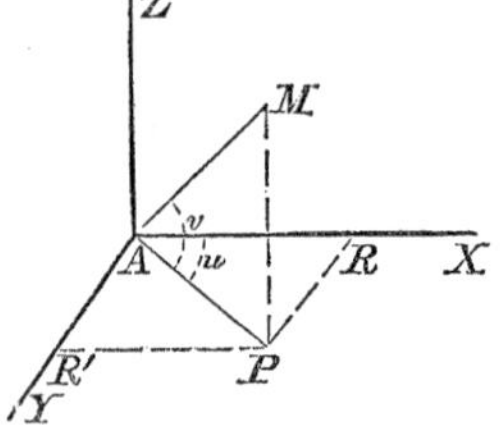

$$AR = x, \quad AR' = y, \quad MP = z,$$

and to the fixed plane XY, the line AX and the point A, by the polar co-ordinates, Art. (43),

$$AM = r, \qquad MAP = v, \qquad RAP = u.$$

The right angled triangles ARP and MPA, give

$$AR = AP \cos u, \qquad RP = AP \sin u,$$

$$MP = r \sin v, \qquad AP = r \cos v.$$

Substituting the value of AP, the first three equations give

$$x = r \cos v \cos u, \quad y = r \cos v \sin u, \quad z = r \sin v \ldots\ldots(1),$$

which are *formulas for passing from a system of rectangular co-ordinates* to a system of polar co-ordinates, in space.

From the last of the above equations, we have

$$r = \frac{z}{\sin v},$$

and since z and the sin v will always have the same sign, *the radius vector will always be positive.*

The equations of the radius vector in any one of its positions, will be of the form, Art. (45),

$$x = az, \qquad y = bz \ldots\ldots(2),$$

whence

$$a = \frac{x}{z}, \qquad b = \frac{y}{z}.$$

Substituting the values of x, y and z, taken from formulas (1), we have

$$a = \cot v \cos u, \qquad b = \cot v \sin u,$$

and these, in equations (2), give

$$x = \cot v \cos uz, \qquad y = \cot v \sin uz,$$

which will be given, when v and u are known.

OF THE CYLINDER.

74. *A cylindrical surface* or *cylinder*, may be generated by moving a straight line, so as to touch a given curve and have all of its positions parallel to its first position.

The moving line is called *the generatrix;* and the given curve *the directrix of the cylinder.*

The different positions of the generatrix are called *elements of the surface.*

The curve of intersection of the cylinder, by any plane, may be regarded as *the base* of the cylinder; and when the elements are perpendicular to the base, the surface is *a right cylinder.*

75. If the directrix of the cylinder is a plane curve, its plane may be taken for the co-ordinate plane XY, and its equation may be represented, generally, by

$$f(x, y) = 0 \ldots\ldots\ldots\ldots (1),$$

which is read, a function of x and y equal to zero; the first member being a symbol to indicate an expression containing x, y and constants; or that x and y are so connected that one can not vary without the other.

Let

$$x = az + \alpha, \qquad y = bz + \beta,$$

be the equations of a right line which is to be moved so as to generate the surface. Since the different positions of this generatrix are parallel, a and b remain constant, while, as the line is moved from one position to another, α and β must change. But α and β are the co-ordinates of the point in which the generatrix pierces the plane XY, Art. (46), and since this point must be on the directrix CD, the values of α and β, when substituted for x and y, must satisfy equation (1). These values are

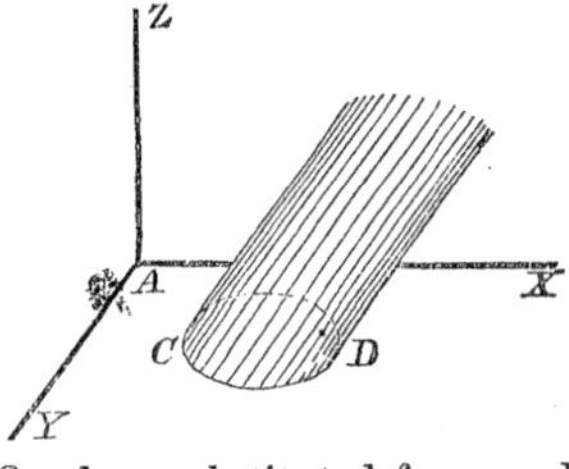

$$\alpha = x - az, \qquad \beta = y - bz,$$

and when substituted in equation (1), give

$$f(x - az, \; y - bz) = 0,$$

an equation expressing the relation between the co-ordinates of the different points of the generatrix in all of its positions. It is, therefore, *the general equation of a cylinder*, of which the directrix may be regarded as the base.

In order then, to obtain the particular equation of a cylinder, whose directrix is given, we have simply *to substitute, for x and y in the equation of the directrix, the expressions*

$$x - az, \qquad y - bz.$$

76. If the directrix is a circle, whose equation is

$$x^2 + y^2 = R^2,$$

the origin being at the centre, we have, by making the substitutions above referred to,

$$(x - az)^2 + (y - bz)^2 = R^2 \dots\dots\dots\dots (1),$$

the equation of an oblique cylinder with a circular base.

If this cylinder be intersected by a plane parallel to XY, the equation of which, Art. (62), is

$$z = g, \qquad x \text{ and } y \textit{ indeterminate},$$

we have, by combining the equations, Art. (62),

$$(x - ag)^2 + (y - bg)^2 = R^2,$$

for the projection of the curve of intersection on XY. But this is evidently the equation of a circle, whose radius is R, Art. (34), and therefore equal to the base. But since this intersection is parallel to the plane XY, its projection is evidently equal to the line itself. We therefore conclude, that if a cylinder, with a circular base, be intersected by a plane parallel to the base, the intersection will be a circle equal to the base.

If a and b are equal to 0, the generatrix becomes parallel to the axis of Z, or perpendicular to the base, the cylinder becomes right, and equation (1) reduces to

$$x^2 + y^2 = R^2,$$

the same as the equation of the base, *z being indeterminate.*

OF THE CONE.

77. *A conical surface, or cone*, may be generated by moving a straight line, so as, continually, to pass through a fixed point and touch a given curve.

The fixed point is *the vertex* of the cone, and the parts of the surface separated by the vertex are called *nappes*.

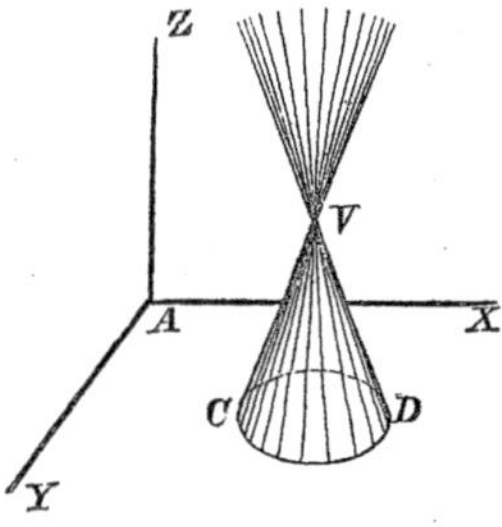

The intersection of the cone by any plane may be regarded as its *base*. The right line drawn from the vertex to the centre of the base is *the axis* of the cone, and if this axis is perpendicular to the plane of the base, the cone is *a right cone*.

78. If the directrix of the cone is a plane curve, its plane may be taken as the co-ordinate plane XY, and its equation be represented as in article (75), by

$$f(x,y) = 0 \text{............} (1).$$

If x', y' and z' are the co-ordinates of the fixed point, or vertex, the equations of the generatrix will be, Art. (50),

$$x - x' = a(z - z'), \qquad y - y' = b(z - z') \text{............} (2),$$

in which a and b change as the generatrix is moved from one position to another. These equations may be put under the form,

$$x = az + (x' - az'), \qquad y = bz + (y' - bz'),$$

in which the absolute terms,

$$x' - az', \qquad y' - bz',$$

are the co-ordinates of the point, in which the line pierces the plane XY, Art. (46), and since this point is on the directrix, whatever be the position of the generatrix, these values, when substituted for x and y in equation (1), must satisfy it, and give

$$f(x' - az', \ y' - bz') = 0.$$

Substituting in this equation, the values of a and b, in terms of x, y and z, deduced from equations (2),

$$a = \frac{x - x'}{z - z'}, \qquad b = \frac{y - y'}{z - z'},$$

we have after reduction,

$$f\left(\frac{x'z - z'x}{z - z'}, \quad \frac{y'z - z'y}{z - z'}\right) = 0 \ldots\ldots\ldots\ldots(3),$$

an equation expressing the relation between x, y and z, for all positions of the generatrix. It is, therefore, *the general equation of a cone*, of which the directrix may be regarded as the base.

In order then to obtain the particular equation of a cone, whose directrix is given, we have simply *to substitute for x and y, in the equation of the directrix, the expressions*,

$$\frac{x'z - z'x}{z - z'}, \qquad \frac{y'z - z'y}{z - z'}.$$

79. If the directrix is a circle, whose equation is

$$x^2 + y^2 = R^2,$$

we have, by making the substitutions above referred to,

$$\left(\frac{x'z - z'x}{z - z'}\right)^2 + \left(\frac{y'z - z'y}{z - z'}\right)^2 = R^2,$$

or

$$(x'z - z'x)^2 + (y'z - z'y)^2 = R^2 (z - z')^2 \ldots\ldots\ldots\ldots(1),$$

for *the equation of an oblique cone with a circular base.*

If this cone be intersected by a plane parallel to XY, the equation of which, Art. (62), is

$$z = g, \qquad x \textit{ and } y \textit{ indeterminate},$$

we have, by combining the equations,

$$(x'g - z'x)^2 + (y'g - z'y)^2 = R^2 (g - z')^2,$$

for the projection of the curve of intersection on XY. By dividing both members by z'^2, this equation may be put under the form

$$\left(\frac{x'g}{z'} - x\right)^2 + \left(\frac{y'g}{z'} - y\right)^2 = \frac{R^2}{z'^2} (g - z')^2,$$

which is the equation of a circle, the co-ordinates of whose centre are $\frac{x'g}{z'}$ and $\frac{y g}{z'}$, and the radius, the square root of the second member, Art. (34). This projection being equal to the curve itself, we conclude, that if a cone, with a circular base, be intersected by a plane parallel to the base, *the intersection will be a circle.* The radius of this circle will decrease as g increases, until $g = z'$, when the radius becomes 0 and the equation takes the form

$$(x' - x)^2 + (y' - y)^2 = 0,$$

which can only be satisfied, Art. (49), by making

$$x = x', \qquad y = y',$$

and the circle becomes a point.

80. If

$$x' = 0, \qquad y' = 0, \qquad z' = h,$$

the vertex of the cone is on the axis of Z, at a distance, from the origin, represented by h; the cone becomes right, and equation (1), of the preceding article, becomes

$$(x^2 + y^2)\, h^2 = R^2\,(z - h)^2,$$

or

$$(x^2 + y^2)\frac{h^2}{R^2} = (z - h)^2 \ldots\ldots\ldots\ldots (1).$$

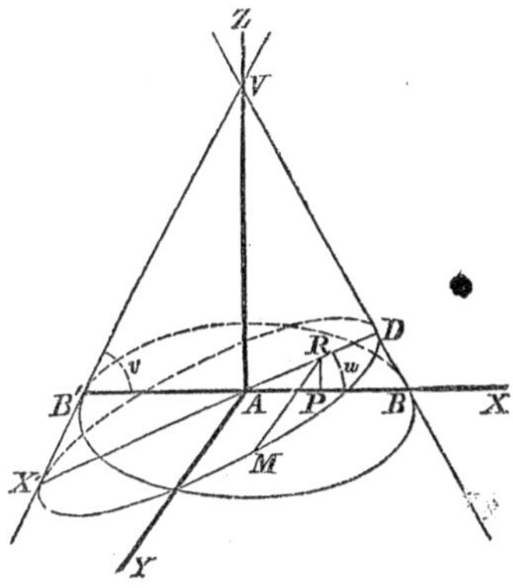

If the angle, made by the elements of the cone with the plane of the base, be denoted by v, we have in the right angled triangle VAB′,

$$\text{tang AB'V} = \frac{\text{AV}}{\text{AB'}}, \qquad \text{or} \qquad \text{tang } v = \frac{h}{R},$$

and equation (1) becomes

$$(x^2 + y^2)\,\text{tang}^2 v = (z - h)^2 \ldots\ldots\ldots\ldots (2),$$

for *the equation of a right cone with a circular base.*

81. Through the axis of Y, in the figure of the preceding article, let a plane be passed intersecting the cone. This plane being perpendicular to the plane XZ, its equation will be the same as that of its trace on XZ, y being indeterminate, Art. (56). Let the angle,

which this plane makes with XY, be denoted by u, the equation of its trace, AR, will be, Art. (24),

$$z = \text{tang } ux.$$

The equations of the curve of intersection of the plane and cone may now be found, as in article (62). But as the different curves, obtained by changing the position of the cutting plane, form a class possessing very remarkable properties, the discussion of which is much simplified by referring the intersection to lines in its own plane, the latter method is chosen.

Let us then take the right lines AX′ and AY, as a new system of rectangular co-ördinate axes, and let us estimate the positive values of x' from A to X′, and the positive values of y' from A to Y.

Let M be any point of the curve of intersection. Its co-ordinates, referred to the primitive planes, are

$$x = \text{AP}, \qquad y = \text{MR}, \qquad z = \text{RP},$$

and referred to the new axes, AX′ and AY,

$$-x' = \text{AR}, \qquad y' = \text{MR}.$$

From the right angled triangle APR, we have

$$\text{AP} = \text{AR} \cos u, \qquad \text{RP} = \text{AR} \sin u,$$

or

$$x = -x' \cos u, \qquad z = -x' \sin u.$$

We have also

$$y = y'.$$

If these values of x, y and z be substituted in equation (2) of the preceding article, the result expressing a relation between x' and y' for points common to the plane and cone only, will be the equation of the intersection. Making the substitution, we obtain

$$(x'^2 \cos^2 u + y'^2) \text{tang}^2 v = (-x' \sin u - h)^2,$$

or performing the operation indicated in the second member, and transposing,

$$y'^2 \text{ tang}^2 v = x'^2 \sin^2 u - x'^2 \cos^2 u \text{ tang}^2 v + 2x'h \sin u + h^2,$$

or recollecting that

$$\sin^2 u = \cos^2 u \text{ tang}^2 u,$$

and omitting the dashes of the variables,

$$y^2 \text{ tang}^2 v = x^2 \cos^2 u (\text{tang}^2 u - \text{tang}^2 v) + 2xh \sin u + h^2 \ldots (1),$$

for *the equation of the line of intersection of a plane and right cone with a circular base.*

In this equation, h may now be regarded as the distance from the vertex of the cone to the point in which the plane cuts the axis.

82. If in the above equation, v *remaining the same*, all values be assigned to u from 0 to 90°, and all values to h, from 0 to infinity, it will represent, in succession, every line which it is possible to cut, from a given right cone with a circular base, by a plane.

There are three distinct cases.

First, when

$$u = v, \qquad \text{or} \qquad \text{tang } u = \text{tang } v.$$

In this case, the cutting plane makes the same angle with the base that the elements do, or is parallel to one of the elements, and since

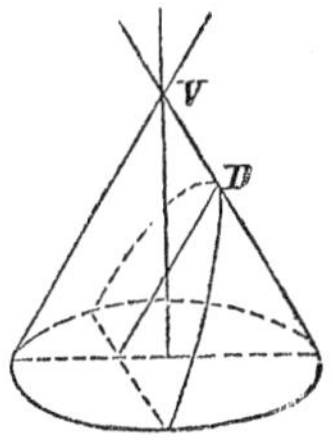

$$\text{tang}^2 u = \text{tang}^2 v,$$

the coefficient of x^2 becomes 0, the equation reduces to

$$y^2 \text{ tang}^2 v = 2xh \sin u + h^2,$$

and the curve represented by it is called *a Parabola.*

If in this equation, $h = 0$, the cutting plane passes through the vertex, and the equation reduces to

$$y^2 \text{ tang}^2 v = 0,$$

which can only be satisfied by making

$$y = 0,$$

which, since x is indeterminate, is the equation of the axis of X, Art. (21). *A right line is therefore regarded as a particular case of the parabola.*

Second, when

$$u < v, \qquad \text{or} \qquad \text{tang } u < \text{tang } v.$$

In this case, the cutting plane makes a less angle with the base than the elements do, or is parallel to none of the elements, see figure of Art. (80); and since,

$$\text{tang}^2 u < \text{tang}^2 v,$$

the coefficient of x^2 is essentially *negative* and the curve represented by the equation is called *an Ellipse.*

If in this case $u = 0$, the cutting plane is parallel to the base,

$$\cos u = 1, \qquad \sin u = 0, \qquad \text{tang } u = 0,$$

and the equation reduces to

$$y^2 \text{ tang}^2 v = - x^2 \text{ tang}^2 v + h^2,$$

or dividing by $\text{tang}^2 v$ and transposing

$$y^2 + x^2 = \frac{h^2}{\text{tang}^2 v},$$

which is the equation of a circle, Art. (35).

If $h = 0$, u being still less than v, the plane passes through the vertex, and the equation reduces to

$$y^2 \text{ tang}^2 v = x^2 \cos^2 u \; (\text{tang}^2 u - \text{tang}^2 v),$$

the first member of which is essentially *positive* and the second *negative*; it can therefore be satisfied for no values of x and y, except

$$x = 0, \qquad y = 0,$$

which are the equations of the origin of co-ordinates, Art. (16). *A circle and point are therefore regarded as particular cases of the ellipse.*

Third, when

$$u > v, \qquad \text{or} \qquad \text{tang } u > \text{tang } v.$$

In this case, the cutting plane makes a greater angle with the base than the elements do, or is parallel to two of the elements, viz. those cut from the cone by passing a plane through the vertex parallel to the cutting plane, and since

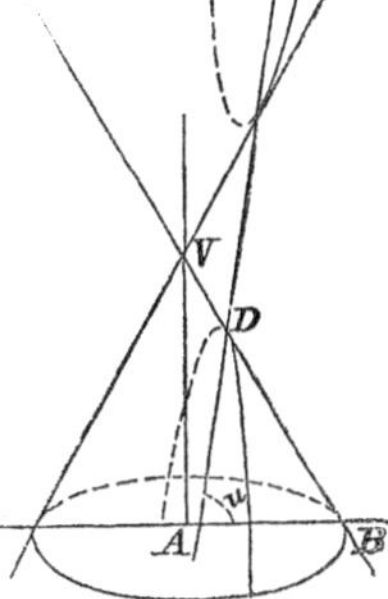

$$\text{tang}^2 u > \text{tang}^2 v,$$

the coefficient of x^2 is *essentially positive*, and the curve represented by the equation is called *an Hyperbola.*

If in this case, $h = 0$, the equation reduces to

$$y^2 \text{ tang}^2 v = x^2 \cos^2 u \; (\text{tang}^2 u - \text{tang}^2 v),$$

both members of which are essentially positive. Dividing by $\text{tang}^2 v$, and placing

$$\frac{\cos^2 u \; (\text{tang}^2 u - \text{tang}^2 v)}{\text{tang}^2 v} = r^2,$$

we obtain

$$y^2 = r^2 x^2, \qquad y = \pm\, rx,$$

which, evidently, represents two right lines intersecting at the origin of co-ordinates, Art. (24), the equations of which are

$$y = + rx, \qquad y = - rx.$$

Two right lines, which intersect, are therefore regarded as a particular case of the hyperbola.

83. Resuming equation (1), Art. (81), dividing by $\tan g^2 v$, and denoting the co-efficient of x^2 by r^2, as above, we have

$$y^2 = r^2x^2 + 2x \frac{h \sin u}{\tan g^2 v} + \frac{h^2}{\tan g^2 v} \ldots\ldots\ldots\ldots (1).$$

Now let us transfer the reference of the points of the curve to a set of parallel co-ordinate axes, having their origin at D, the point in which the curve is cut by the axis of X, [see figure of Art. (80)]. Formulas (2), of Art. (67), become for this case,

$$x = a + x', \qquad y = y',$$

a representing the distance $-$ AD, and b' being equal to 0.

Substituting these values in equation (1), we have

$$y'^2 = r^2x'^2 + 2\left(\frac{h \sin u}{\tan g^2 v} + r^2a\right)x'$$

$$+ r^2a^2 + 2 \frac{h \sin u}{\tan g^2 v} a + \frac{h^2}{\tan g^2 v}.$$

The origin of co-ordinates being on the curve, the absolute term

$$r^2a^2 + \frac{2h \sin u}{\tan g^2 v} a + \frac{h^2}{\tan g^2 v},*$$

* NOTE. It should be observed, that by placing the absolute term

$$r^2a^2 + 2 \frac{h \sin u}{\tan g^2 v} a + \frac{h^2}{\tan g^2 v} = 0,$$

must be equal to 0, Art. (38), and the equation, after omitting the dashes and placing

$$\frac{h \sin u}{\text{tang}^2 v} + r^2 a = p,$$

reduces to

$$y^2 = r^2 x^2 + 2px \ldots\ldots\ldots\ldots (2),$$

a general equation, which may represent either of the above named curves; *the parabola* when $r^2 = 0$, *the ellipse* when $r^2 < 0$, and *the hyperbola* when $r^2 > 0$.

OF THE PARABOLA.

84. If $r^2 = 0$, equation (2) of the preceding article, becomes

$$y^2 = 2px \ldots\ldots\ldots\ldots (1).$$

This equation being of the second degree, the line represented by it is of the second order, Art. (33), and $2p$ being the only con-

---

we have an equation of the second degree, and therefore two values of a, which will fulfil the required condition. Solving the equation, substituting the value of r^2 and reducing, we find

$$a = -\frac{h\,(\text{tang}\ u \mp \text{tang}\ v)}{\cos\ u\,(\text{tang}^2 u - \text{tang}^2 v)}.$$

In the parabola, u being equal to v, the first value reduces to $\frac{0}{0}$, and the second, to infinity, but by striking out the common factor, $\text{tang}\ u - \text{tang}\ v$, the first value becomes finite and negative, as it should be to give the point D.

In the ellipse, the first value is negative, the other positive, the negative value being used.

In the hyperbola, both values are negative, the one which is numerically the least being used.

stant, the line is given when $2p$ is given, Art. (23). This constant is called *the parameter* of the parabola, and since from equation (1), we may deduce the proportion

$$x : y :: y : 2p,$$

we say, *the parameter is a third proportional to the abscissa and ordinate of any point of the curve.*

85. If equation (1), of the preceding article, be solved with reference to y, we have

$$y = \pm \sqrt{2px}.$$

For *every positive value of* x, there will be two corresponding real values of y; hence, the curve is continuous and extends from the origin, A, to infinity, in the direction of the positive abscissas; and since these values of y are equal with contrary signs, it follows that for each assumed abscissa, as AP, there will be two corresponding points of the curve, one above and the other below the axis of X, at equal distances, and the two values of y taken together will form a chord, as MM′, which will be bisected by the axis of X; hence, *the curve is symmetrical with regard to the axis of* X.

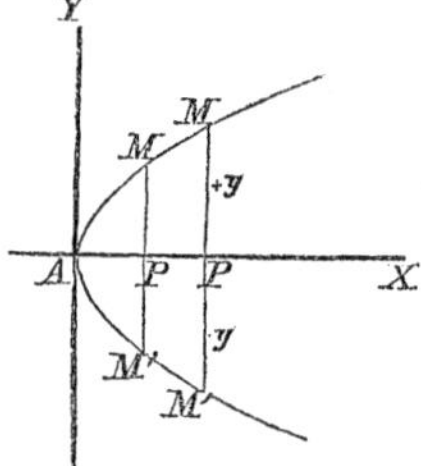

The line AX is called *the axis of the parabola*, and the point A, in which it intersects the curve, is called *the vertex*; and, in general, *any straight line, which bisects a system of chords perpendicular to it, is an axis* of the curve in which the chords are drawn.

If $x = 0$, we have

$$y = \pm 0,$$

which proves that the curve is tangent to the axis of Y, at the origin, Art. (34).

If x *is negative*, the values of y are imaginary; hence, there is no point of the curve on the left of the axis of Y,

If $y = 0$, we have

$$x = 0,$$

and the curve cuts the axis of X in one point only, at the origin.

86. The curve may be constructed by points from its equation as in Art. (22). This is done geometrically, thus: Let AX and AY be two co-ordinate axes at right angles. Lay off from the origin in the direction of the negative abscissas AP′ = $2p$, and take any positive abscissa, as AP; on the line PP′ as a diameter, describe a circle, and from the points in which it intersects the axis of Y, draw lines parallel to the axis of X until they intersect the perpendicular erected to AX, at P. The points of intersection, M and M′, will be points of the curve. For, from a known property of the circle, we have

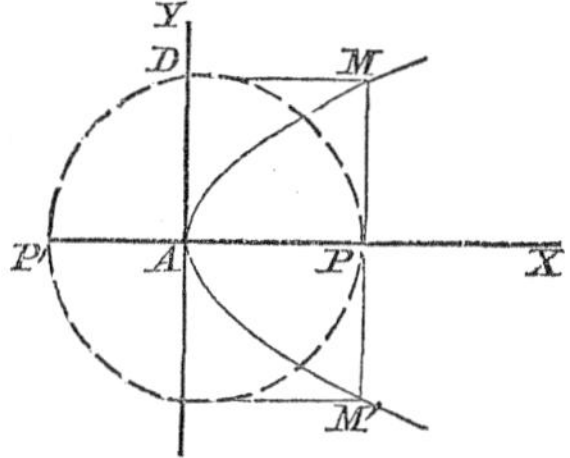

$$\overline{AD}^2 = AP' \times AP = \overline{PM}^2, \qquad \text{or} \qquad y^2 = 2px.$$

87. If a point whose co-ordinates are x and y, is on the curve, we must have the condition, Art. (23),

$$y^2 = 2px, \qquad \text{or} \qquad y^2 - 2px = 0.$$

If the point is without the curve, since its ordinate will be greater than the corresponding ordinate of the curve, we must have

$$y^2 > 2px, \qquad \text{or} \qquad y^2 - 2px > 0.$$

If the point is within the curve,

$$y^2 < 2px, \qquad \text{or} \qquad y^2 - 2px < 0.$$

88. If in equation (1), Art. (84), we make $x = \frac{p}{2}$, we have

$$y^2 = p^2, \qquad y = p, \qquad 2y = 2p.$$

Hence, if a point, as F, be taken on the axis of the parabola, at a distance from the vertex equal to *one fourth of the parameter, the double ordinate, or the chord, perpendicular to the axis at this point, will be equal to the parameter of the curve.*

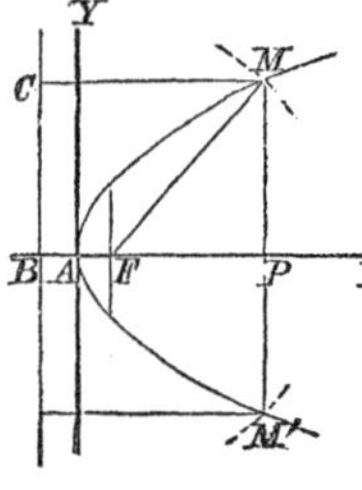

If F be the point and M any point of the curve, the right angled triangle FPM will give

$$\text{FM} = \sqrt{\overline{\text{FP}}^2 + \overline{\text{PM}}^2},$$

or, since

$$\text{FP} = \text{AP} - \text{AF} = x - \frac{p}{2}, \qquad \text{PM} = y,$$

we have

$$\text{FM} = \sqrt{\left(x - \frac{p}{2}\right)^2 + y^2},$$

or squaring $x - \frac{p}{2}$, and substituting for y^2 its value $2px$,

$$\text{FM} = \sqrt{x^2 \mp px + \frac{p^2}{4}} = x + \frac{p}{2}.$$

If from the vertex A, we lay off $AB = -\frac{p}{2}$, and draw BC perpendicular to the axis, we shall have

$$MC = BP = BA + AP = x + \frac{p}{2} = FM.$$

Hence, the distance from any point of the curve to the line BC, is equal to the distance from the same point to the point F.

This remarkable property enables us to define a parabola to be *a curve, such, that each of its points is at the same distance from a given point and a given straight line.*

The given point, F, is called *the focus,* the given line BC, *the directrix,* and a straight line drawn through the focus perpendicular to the directrix, is *the axis of the parabola.*

This property, also, gives another simple method of constructing the curve by points, when the directrix and focus are given. Let BC be the directrix and F the focus. Through F draw FB perpendicular to BC, it will be the axis. At any point of the axis, as P, erect a perpendicular; with the focus F as a centre, and radius BP, describe arcs cutting the perpendicular in M and M′; these will be points of the curve, since

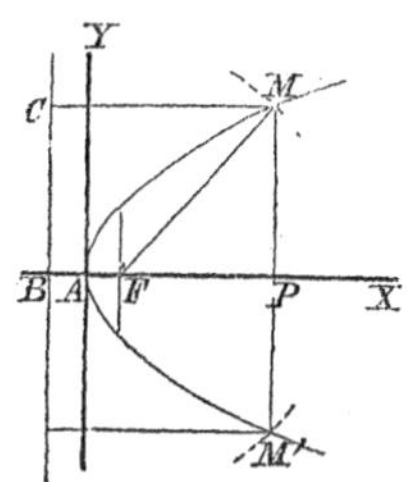

$$FM = BP = MC.$$

The curve may also be constructed by a continuous movement. Place one side DC, of a right angled triangular rule DCE, against the directrix; fasten one end of a string equal in length to the other side EC, at the point E, and the other end at the focus; press a pencil against the string and rule, and as the rule is moved along the directrix, the point of the pencil will describe the parabola; for we always have

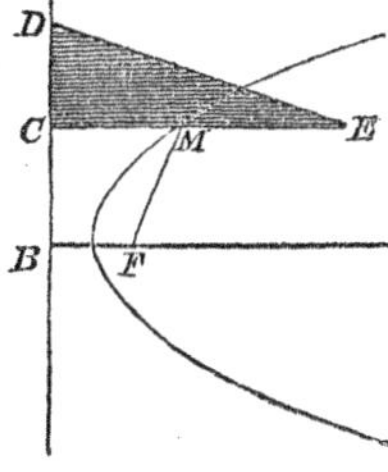

$$FM = MC.$$

89. Let x', y' and x'', y'' be the co-ordinates of any two points of the parabola. Since these are points of the curve, their co-ordinates will satisfy its equation and give the two conditions, Art. (23),

$$y'^2 = 2px', \qquad y''^2 = 2px'',$$

from which, omitting the common multiplier $2p$, we obtain the proportion

$$y'^2 : y''^2 :: x' : x'',$$

that is, *the squares of the ordinates of any two points of the curve are proportional to the corresponding abscissas.*

90. Let x'', y'' be the co-ordinates of any point, as M, on the curve, and through this point conceive any straight line to be drawn; its equation will be of the form, Art. (29),

$$y - y'' = d(x - x'') \text{...........}(1),$$

in which d is undetermined. Since the given point is on the curve, we must have the condition

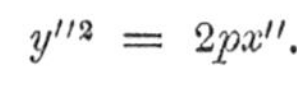

$$y''^2 = 2px''.$$

Subtracting this, member by member, from the equation

$$y^2 = 2px,$$

we have

$$y^2 - y''^2 = 2p(x - x''),$$

or

$$(y + y'')(y - y'') = 2p(x - x''),$$

which is the equation of the parabola, with the condition introduced that the given point shall be on the curve. Combining this with equation (1), by substituting the value of $y - y''$, taken from (1), we obtain

$$(y + y'')\,d(x - x'') = 2p(x - x''),$$

or

$$[(y + y'')\,d - 2p](x - x'') = 0 \ldots\ldots\ldots\ldots(2),$$

in which x and y must represent all the points common to the right line and curve, Art. (27). This equation being of the second degree, there are two such points, and only two; and the equation may be satisfied by placing the factors separately equal to 0. Placing

$$x - x'' = 0, \qquad \text{we have} \qquad x = x'',$$

and this value in (1) gives $y = y''$. The values thus obtained are the co-ordinates of the given point, which is one of the points common to the two lines. By placing the other factor equal to 0, we have

$$(y + y'')d - 2p = 0 \ldots\ldots\ldots\ldots(3),$$

in which y must be the ordinate of the second point of intersection, M′. If now, the right line be revolved about the point M, so as to cause the point M′ to approach M, y in equation (3), becomes nearer and nearer equal to y'', and finally, when the two points coincide, we shall have $y = y''$, the line will be tangent to the curve, and equation (3) reduce to

$$2y''d = 2p, \qquad \text{whence} \qquad d = \frac{p}{y''},$$

which is the value d must have when the assumed line becomes a tangent. Substituting this value of d in (1), we have

$$y - y'' = \frac{p}{y''}(x - x''),$$

or

$$yy' - y''^2 = px - px'',$$

which by the substitution of $2px''$ for y''^2, becomes

$$yy'' = p(x + x'')\dots\dots\dots\dots(4),$$

for *the equation of a tangent line to the parabola at a given point.*

91. If we multiply both members of the last equation by 2, and subtract the result, member by member, from the equation

$$y''^2 = 2px'',$$

we have

$$y''^2 - 2yy'' = - 2px,$$

adding y^2 to both members,

$$y''^2 - 2yy'' + y^2 = y^2 - 2px,$$

or

$$(y'' - y)^2 = y^2 - 2px.$$

The first member being a perfect square, is positive for all values of y except $y = y''$;

$$y^2 - 2px,$$

is therefore positive for all values of y and x, except $y = y''$, $x = x''$, when it will be 0; hence, since x and y are the general co-ordinates of the tangent, *all points of the tangent, except the point of contact, are without the curve,* Art. (87).

92. If in equation (4), Art. (90), we make $y = 0$, we find

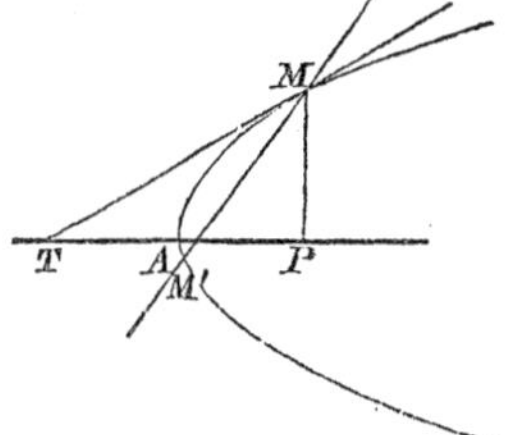

$$0 = p\,(x + x''), \quad \text{or} \quad x = - x'',$$

for the distance AT, to the point in which the tangent cuts the axis; hence, we have

$$\text{PT} = \text{TA} + \text{AP} = 2x''.$$

The distance PT is called *the subtangent*, which, in general, is *the distance from the foot of the ordinate of the point of contact, to the point in which the tangent cuts the axis, to which the ordinate is drawn;* and in the parabola, *is equal to double the abscissa of the point of contact.*

This property gives a simple method of drawing a tangent to a parabola at a given point. Let M be the point. From the vertex lay off, on the axis without the parabola, a distance AT, equal to the abscissa of the given point; draw a right line from the extremity of this distance to the point of contact, it will be the required tangent.

93. If the point M be joined with the focus F, we have, Art. (88),

$$FM = x'' + \frac{p}{2}.$$

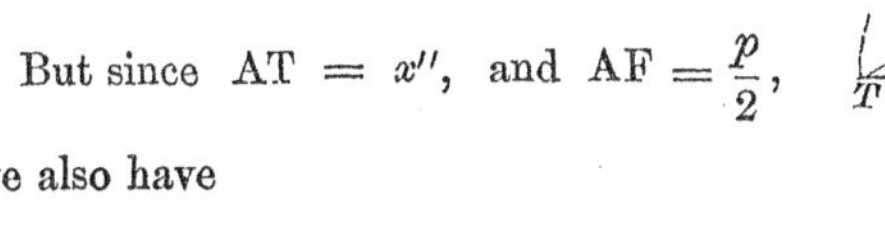

But since $AT = x''$, and $AF = \frac{p}{2}$,

we also have

$$FT = x'' + \frac{p}{2};$$

hence, $FM = FT$, the triangle TFM is isosceles, and the angle

$$FMT = FTM.$$

Hence, *if a right line be drawn from the focus of a parabola to the point of contact of a tangent, this line will make an angle with the tangent equal to that which the tangent makes with the axis.*

This property enables us to make the following constructions.

First. To draw a tangent to the parabola at a given point. Draw a right line from the point, as M, to the focus; with this line as a radius and the focus as a centre, describe an arc cutting the axis, without the curve, in a point, as T; draw a right line from this to the given point, it will be the required tangent, as the triangle MFT will be isosceles.

Or otherwise, thus. Draw a right line through the given point perpendicular to the directrix; join the point C, in which it intersects the directrix, with the focus, and through the given point draw a right line perpendicular to this last line, it will be the tangent. For, since MF = MC, the triangle CMF is isosceles and therefore the angle FMT = CMT; but CMT = MTF; hence,

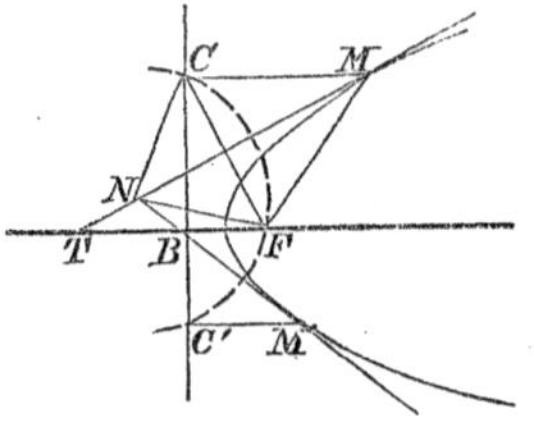

$$\text{FMT} = \text{MTF}.$$

Second. To draw a tangent from a point without the curve, as N. Join the point with the focus; with this distance as a radius, and the given point as a centre, describe an arc cutting the directrix in the points C and C′; through these points, draw lines parallel to the axis, cutting the curve in the points M and M′; join these points with the given point and we shall have the tangents NM and NM′. For, since

$$\text{MF} = \text{MC}, \qquad \text{and} \qquad \text{NF} = \text{NC},$$

the line NM has two of its points equally distant from the points F and C, is therefore perpendicular to FC at its middle point and bisects the angle FMC.

Let the co-ordinates of the given point N, be denoted by x' and y'. Since this point is on the tangent, we must have the equation of condition, Art. (23),

$$y'y'' = p\,(x' + x'')\ldots\ldots\ldots\ldots(1),$$

and since the point of contact is on the parabola, we also have the equation of condition,

$$y''^2 = 2px''.$$

In these equations x'' and y'' are unknown, and since one is of the first and the other of the second degree, their combination will give an equation of the second degree, and there will be two values of x'' and two corresponding of y''.

Combining these equations by substituting the value

$$x'' = \frac{y''^2}{2p},$$

in the first, we obtain

$$y''^2 - 2y'y'' = -2px'\ldots\ldots\ldots\ldots(2),$$

from which we deduce the two values

$$y'' = y' \pm \sqrt{y'^2 - 2px'}.$$

The values of y'' will evidently be real, when

$$y'^2 - 2px' > 0,$$

that is, when the given point is without the curve, Art. (87), and there will be *two tangents*, as appears by the geometrical construction.

The values of y'' will be equal when the point is on the curve and there will be but *one tangent.*

They will be imaginary when the point is within the curve and there will be *no tangent.*

The corresponding values of x'' being found, each set of co-ordinates may be substituted, in succession, in equation (4), Art. (90), and the equations of the two tangents thus determined.

Third. To draw a tangent parallel to a given line as SR. Produce the line until it intersects the axis at S, with the focus as a centre, and the distance FS as a radius describe an arc cutting the given line in R, join this point with the focus, the point M, in which the last line intersects the curve will be the point of contact, through which draw MT parallel to the given line,

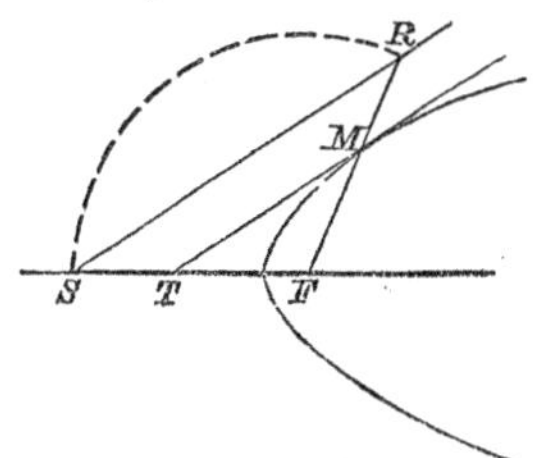

it will be the required tangent. For, since MT is parallel to RS, and FS = FR, we have

$$FM = FT.$$

94. Since the triangle FMT is isosceles, the line FD, drawn perpendicular to the base MT, will pass through its middle point; and since AT = AP, Art. (92), the line AD also passes through the middle point of MT: Hence, *if from the focus of a parabola, a right line be drawn perpendicular to any tangent, it will intersect this tangent, on the tangent at the vertex; and conversely.*

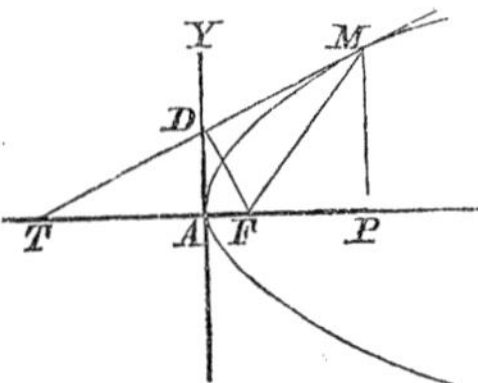

Since the triangle FDT is right angled at D, we have

$$\overline{FD}^2 = AF \times FT,$$

and since AF is constant and FT = FM; the square of the perpendicular FD, will vary as *the first power* of the distance from the focus to the point of contact.

95. If in equation (1), Art. (93),

$$y'y'' = p(x' + x'')............(1),$$

we regard x'' and y'' as variables, it will be the equation of a right line, Art. (25); and since both values of x'' and y'' deduced from equation (2), Art. (93), must satisfy this equation, the line represented by it will pass through both points of contact, and will therefore be the indefinite chord which joins these

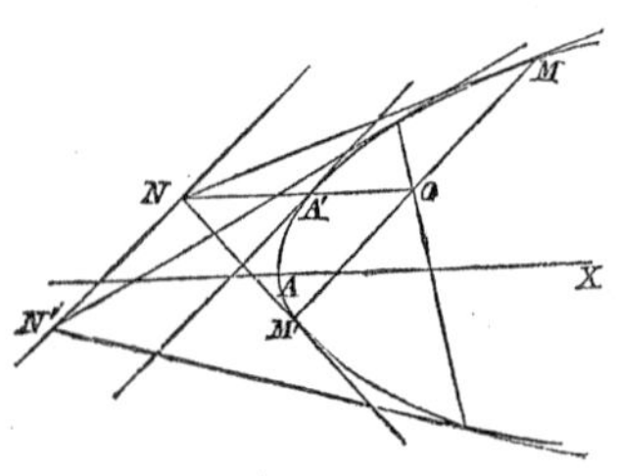

points. If any point as O, be taken upon this chord, its co-ordinates which we will denote by c and d, when substituted for x'' and y'' will satisfy the equation and give the condition

$$y'd = p(x' + c)\ldots\ldots\ldots\ldots(2).$$

Now it is evident, that every set of values for x' and y' which will satisfy this equation, will give a point from which, if two tangents be drawn to the parabola and the points of contact be joined by a chord, this chord will pass through the point O. Hence, if y' and x' be regarded as variables in this equation, it will represent a right line every point of which will fulfil the above condition.

This line is called *the polar line* of the point O, which is called *the pole.*

If through the point O, a line be drawn parallel to the axis AX, the ordinate of the point in which it intersects the curve will be equal to d, and the equation of a tangent to the parabola, at this point, will be, Art. (90),

$$yd = p(x + x''),$$

and this tangent is evidently parallel to the line represented by equation (2), that is to the polar line, Art. (28),

If the line OA′ be further produced till it intersects the polar line in N, the ordinate of this point will be d, which substituted for y' in equation (1), will give

$$y''d = p(x + x''),$$

for the equation of the chord corresponding to this point N, and this is parallel both to the polar line and tangent.

These properties give the following constructions :

1. The pole being given, to construct the corresponding polar line.

Through the pole draw a line parallel to the axis of the parabola ; at the point in which this intersects the curve, draw a tangent; through the pole draw a chord parallel to this tangent, and at its extremity draw another tangent; through the point in which this intersects the line first drawn, draw a line parallel to the chord, it will be the polar line.

2. The polar line being given, to construct the pole.

Draw a tangent parallel to the polar line; through the point of contact draw a line parallel to the axis; from the point in which this intersects the polar line, draw another tangent; through the point of contact thus determined, draw a chord parallel to the polar line, it will intersect the line parallel to the axis in the required pole.

96. If the focus be taken as the pole, the co-ordinates of which are

$$d = 0, \qquad c = \frac{p}{2},$$

equation (2) of the preceding article reduces to

$$0 = p\left(x' + \frac{p}{2}\right), \qquad \text{or} \qquad x' = -\frac{p}{2},$$

y' being indeterminate, which is the equation of the directrix, Art. (21). The directrix is then the polar line of the focus. Hence, *if any chord be drawn through the focus of a parabola and two tangents be drawn at its extremities, these tangents will intersect on the directrix.*

97. If in the general equation of a right line passing through a given point, Art. (29), we substitute for x' and y', the co-ordinates of the focus, we shall have

$$y = a\left(x - \frac{p}{2}\right) \text{..........(1)},$$

for the equation of any chord passing through the focus. Combining this with the equation of the parabola, $y^2 = 2px$, by substituting the value $x = \frac{y^2}{2p}$, we have

$$y = a\left(\frac{y^2}{2p} - \frac{p}{2}\right), \qquad \text{or} \qquad y^2 - \frac{2py}{a} = p^2.$$

Denoting the two roots of this equation by y' and y'', we have from a well known principle of Algebra,

$$y'y'' = -p^2,$$

and if d and d' denote the tangents of the angles made with the axis, by two tangents drawn at the extremities of this chord, we have, Art. (90),

$$d = \frac{p}{y'}, \qquad d' = \frac{p}{y''};$$

whence,

$$dd' = \frac{p^2}{y'y''},$$

or substituting for $y'y''$ the above value,

$$dd' = -1, \qquad \text{or} \qquad dd' + 1 = 0.$$

Hence, Art. (28), *if at the extremities of a chord passing through the focus of a parabola, two tangents be drawn, they will be perpendicular to each other, and intersect on the directrix*, Art. (96).

And conversely, *if two tangents to the parabola are perpendicular to each other, the chord joining their points of contact will pass through the focus.* For, let S′M and S′M″ be the two tangents. If the chord MM″ does not pass through the focus; through the focus and the point M, draw MM′; at M′ draw the tangent M′S. From what precedes, it must be perpendicular to MS′; hence, SM′

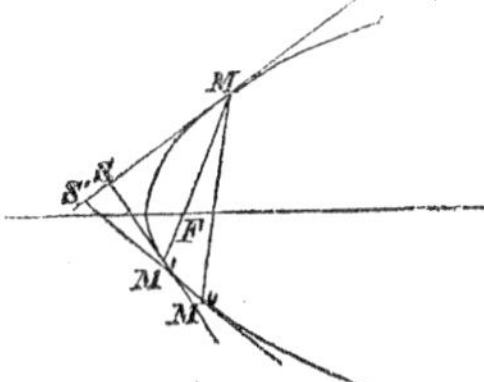

and S′M″ must be parallel. But since the tangent of the angle which a tangent to the parabola makes with the axis is $\frac{p}{y''}$, Art. (90), no two tangents can be parallel, for no two points have equal ordinates. It is then absurd to suppose that MM″ does not pass through F.

98. Through the point of contact of a tangent, let any other straight line be drawn, its equation will be of the form, Art. (29),

$$y - y'' = d'(x - x'') \ldots\ldots\ldots\ldots(1).$$

If this line is perpendicular to the tangent, we must have, Art. (28),

$$dd' + 1 = 0, \quad \text{or} \quad d' = -\frac{1}{d},$$

But, Art. (90),

$$d = \frac{p}{y''};$$

whence,

$$d' = -\frac{y''}{p}.$$

Substituting this in equation (1), we have

$$y - y'' = -\frac{y''}{p}(x - x'') \ldots\ldots\ldots\ldots(2),$$

for the equation of a straight line perpendicular to the tangent at the point of contact. This line is called *a normal.*

If we make $y = 0$, in equation (2), we have

$$-y'' = -\frac{y''}{p}(x - x''),$$

or

$$x - x'' = p,$$

in which, x is the distance AR from the origin to the point in which the normal intersects the axis, and

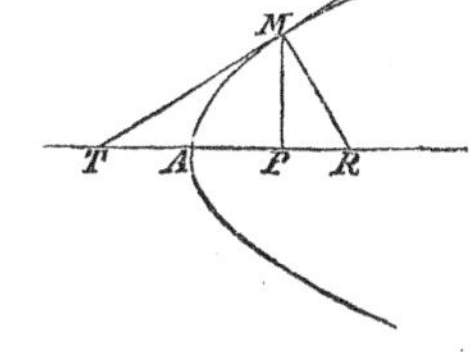

$$x - x'' = \text{AR} - \text{AP} = \text{PR} = p.$$

The distance PR, from the foot of the ordinate of the point of contact, to the point in which the normal cuts the axis, is called *the subnormal.* Hence, the subnormal in the parabola *is constant and equal to half the parameter of the curve.*

This property enables us to construct a tangent at a given point. Draw the ordinate of the point; from its foot lay off a distance equal to one half the parameter; join the extremity of this distance with the given point, through which draw a perpendicular to the last line, it will be the required tangent.

OF THE PARABOLA REFERRED TO OBLIQUE CO-ORDINATE AXES.

99. It was observed in Art. (71), that two classes of propositions might arise in the transformation of co-ordinates. As an example of the second class, let it now be proposed to ascertain if there are any other co-ordinate axes, to which if the parabola be referred, its equation will be of the same form as when referred, to its axis and the tangent at its vertex.

For this purpose, let us take the general formulas (3), Art. (67),

$$x = a + x' \cos \alpha + y' \cos \alpha',$$

$$y = b + x' \sin \alpha + y' \sin \alpha',$$

and substitute the values of x and y in the equation

$$y^2 = 2px \ldots\ldots\ldots\ldots (1).$$

We thus obtain

$$b^2 + 2bx' \sin \alpha + x'^2 \sin^2 \alpha + 2by' \sin \alpha' + 2x'y' \sin \alpha \sin \alpha'$$

$$+ y'^2 \sin^2 \alpha' = 2pa + 2px' \cos \alpha + 2py' \cos \alpha',$$

or transposing, arranging and omitting the dashes of the variables,

$$y^2 \sin^2 \alpha' + x^2 \sin^2 \alpha + 2xy \sin \alpha \sin \alpha'$$

$$+ 2(b \sin \alpha' - p \cos \alpha')y + b^2 - 2pa = 2(p \cos \alpha - b \sin \alpha)x \ldots (2),$$

which is the equation of the parabola referred to any oblique axes. In order that this equation shall be of the same form as equation (1), the absolute term, in the first member, and the terms containing x^2, xy, and y, must disappear, which requires that

$$b^2 - 2pa = 0 \ldots\ldots\ldots\ldots (3);$$

$$\sin^2 \alpha = 0 \ldots\ldots\ldots\ldots (4);$$

$$\sin \alpha \sin \alpha' = 0 \ldots\ldots\ldots\ldots (5);$$

$$b \sin \alpha' - p \cos \alpha' = 0 \ldots\ldots\ldots\ldots (6).$$

These equations contain four arbitrary constants, it is therefore possible to assign such values to the constants as to satisfy the four equations, and thus reduce equation (2) to the proposed form.

Equation (3) is the equation of condition that *the new origin shall be on the curve*, Art. (87).

Equation (4) can only be satisfied by $\alpha = 0$, or $= 180°$; hence *the new axis of* X *must be parallel to the axis of the curve.*

Equation (5) is satisfied by $\sin \alpha = 0$, without introducing any new condition.

Equation (6) can be put under the form

$$\frac{\sin \alpha'}{\cos \alpha'} = \text{tang } \alpha' = \frac{p}{b},$$

and therefore, Art. (90), expresses the condition that *the new axis of* Y *shall be tangent to the curve.*

Since we have thus far introduced but three independent conditions, and since a, b and α' are still undetermined, we may assign a value at pleasure to either of them, whence the other two will become known, and *an infinite number of sets of co-ordinate axes* be thus determined, which will fulfil the required condition, each of which will be subjected to the three conditions expressed by equations (3), (4) and (6).

Substituting the above conditions in equation (2), and observing that, since $\sin \alpha = 0$, $\cos \alpha = 1$, we have

$$y^2 \sin^2 \alpha' = 2px, \qquad \text{or} \qquad y^2 = \frac{2p}{\sin^2 \alpha'} x;$$

or, denoting the coefficient of x by $2p'$

$$y^2 = 2p'x \ldots\ldots\ldots\ldots (7),$$

an equation of the same form as (1).

100. Solving the last equation with reference to y, we find

$$y = \pm \sqrt{2p'x},$$

and we see, as in Art. (85), that every positive value of x, gives two real values of y, equal with contrary signs, and that these two values taken together form a chord, as MM′, parallel to the axis of Y, which chord is bisected by the axis of X, at P. The line A′X is therefore called a diameter of the parabola; and, in general, *any straight line which bisects a system of parallel chords is a diameter*, of the curve in which the chords are drawn. The points in which a diameter intersects the curve are called *the vertices of the diameter.*

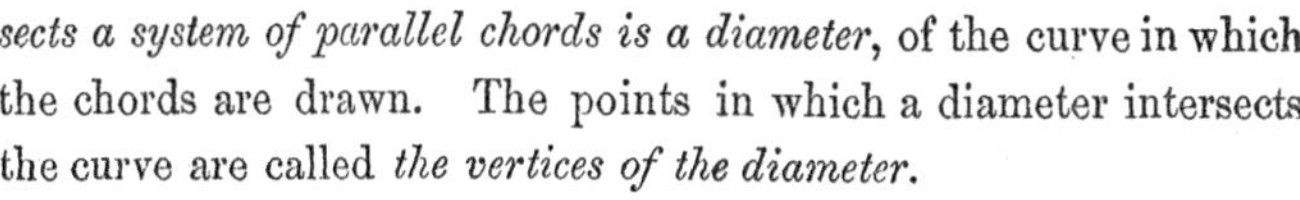

Since condition (4) of the preceding article requires the new axis of X to be parallel to the axis of the curve, it follows that all *the diameters of the parabola are parallel to each other.*

Since condition (6) requires the new axis of Y to be tangent to the curve at the origin, it also follows that *each diameter bisects a system of chords parallel to the tangent at its vertex.*

If the parabola is given, traced upon paper, a diameter may be found by drawing any two parallel chords as MM′ and bisecting them by a straight line as PP; this line will be a diameter.

Draw any two chords perpendicular to this diameter and bisect them by a straight line, this will be the axis, Art. (85). At the vertex of the first diameter, A′, draw a line parallel to the chords which it bisects, it will be a tangent to the curve. At the vertex, A, of the parabola, draw a line perpendicular to the axis, it will also be a tangent.

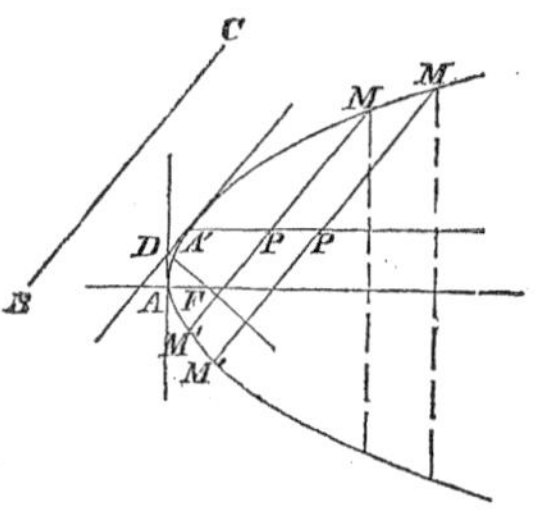

At the point D, where these tangents intersect, draw a perpendicular to the first, it will intersect the axis in the focus F, Art. (94).

The property, that each diameter bisects a system of chords parallel to a tangent at its vertex, suggests the following construction for drawing a tangent parallel to a given line, as BC. Draw two chords parallel to the given line; bisect them by a straight line PP, and at the point A′, where this intersects the curve, draw a line parallel to the given line, it will be the required tangent.

101. The coefficient $2p'$ in equation (7), Art. (99), is called the parameter of the diameter A′X, and, as in Art. (84), is a third proportional to any ordinate and its corresponding abscissa.

If we represent the distance FA′ by r, and recollect that the angle FA′D = FTD is denoted by α', Art. (99), we shall have in the right angled triangle FDA′

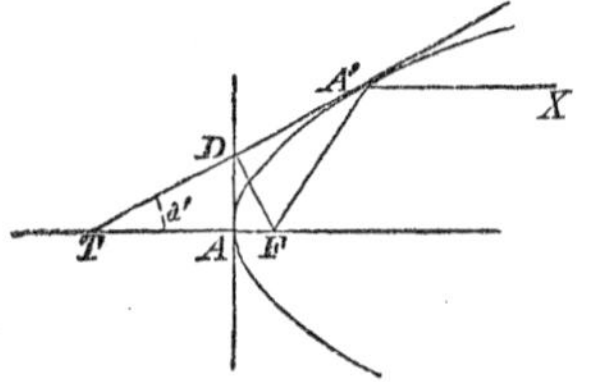

$$FD = r \sin \alpha',$$

or

$$\overline{FD}^2 = r^2 \sin^2 \alpha'.$$

But we also have, Art. (94).

$$\overline{FD}^2 = FA \times FA', \quad \text{or} \quad \overline{FD}^2 = \frac{p}{2} r.$$

Equating these two values of $\overline{FD}^2$, we have

$$r^2 \sin^2 \alpha' = \frac{p}{2} r; \quad \text{whence} \quad \sin^2 \alpha' = \frac{p}{2r}.$$

Substituting this value of $\sin^2 \alpha'$, in the expression, Art. (99),

$$2p' = \frac{2p}{\sin^2 \alpha'},$$

it reduces to

$$2p' = 4r$$

that is, *the parameter of any diameter of the parabola, is equal to four times the distance from the vertex of the diameter to the focus.*

102. Let x'' and y'' be the co-ordinates of any point of the parabola referred to the diameter A′X and the tangent A′Y. The equation of a right line passing through this point will be

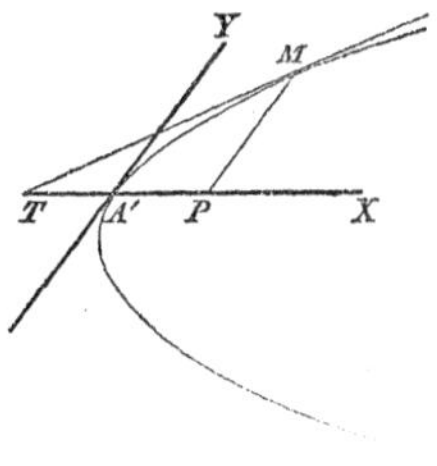

$$y - y'' = d(x - x''),$$

in which d will represent the ratio of the sines of the angles which the line makes with the co-ordinate axes, Art. (20).

By a process identical with that pursued in Art. (90), we can find the value of d, when the line becomes a tangent, and thus deduce the equation of the tangent,

$$yy'' = p'(x + x'').$$

By making $y = 0$, we find

$$x = -x'' = \text{A'T};$$

hence as in Art. (92), the subtangent PT, *is equal to twice the abscissa of the point of contact.* And a tangent may be drawn at a given point by drawing the ordinate MP, of the point, parallel to the axis A′Y, laying off the distance A′T equal to the abscissa A′P, and joining the extremity of this distance with the given point.

103. Let AM be an arc of a parabola, in which inscribe any polygon, as AM′......MP. At the points M, M′, &c., draw the tangents MT, M′T′, &c. Through the middle points of the chords MM′, &c., draw the diameters RS, R″S′, &c., and draw the ordinates MP, M′P′, &c. It is evident that for each chord there will be a trapezoid, as MM′P′P, within the parabola, and a corresponding triangle, as OTT′, without.

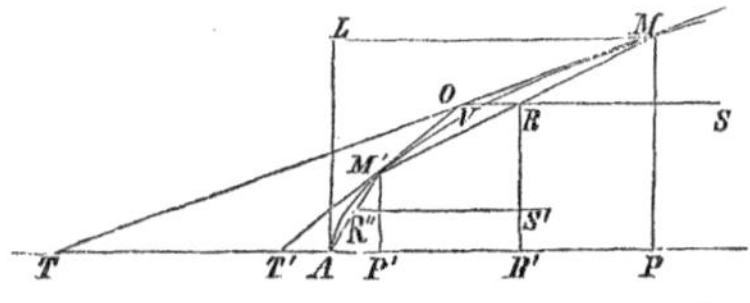

Since the points of contact M and M′, when referred to the diameter RS and tangent line at its vertex, have the same abscissa VR, the subtangent will be the same for each, Art. (102), and the two tangents MO and M′O will intersect the diameter VS, at the same point O; hence the altitude of the triangle OTT′ will be equal to the line RR′, drawn through the middle points of the two inclined sides of the trapezoid P′M′MP; and since,

$$\text{AP} = \text{AT}, \quad \text{and} \quad \text{AP}' = \text{AT}',$$

we have

$$\text{PP}' = \text{TT}';$$

hence, the area of the trapezoid, which is measured by

$$RR' \times PP',$$

is double the area of the triangle, which is measured by

$$\frac{1}{2}RR' \times TT';$$

and so for each trapezoid and corresponding triangle, and the sum of all the interior trapezoids will be double the sum of the corresponding triangles.

If now, the number of sides of the polygon be increased indefinitely, the sum of the trapezoids will be the same as the curvilinear area AM′MP, and the sum of the triangles the same as the exterior area TMM′A; hence the first area is double the second. But the sum of these two areas is equal to the area of the triangle MTP, therefore

$$AM'MP = \frac{2}{3}MTP.$$

But since TP = 2AP, we have

$$\text{triangle } MTP = \text{rectangle } ALMP.$$

Therefore, the area of a portion of the parabola *is equal to two thirds of the rectangle described on the ordinate and abscissa of the extreme point.*

OF THE POLAR EQUATION OF THE PARABOLA.

104. Let us resume the equation

$$y^2 = 2px,$$

and substitute for y and x, their values taken from the formulas (2) of Art. (69);

$$x = a' + r \cos v, \qquad y = b' + r \sin v.$$

We thus obtain

$$b'^2 + 2b'r \sin v + r^2 \sin^2 v = 2p(a' + r \cos v),$$

or transposing and arranging,

$$r^2 \sin^2 v + 2(b' \sin v - p \cos v) r + b'^2 - 2pa' = 0 \ldots\ldots(1),$$

which is *the general polar equation of the parabola*, Art. (69).

By assigning particular values to a' and b', the pole may be placed at any point in the plane of the curve.

First. If it be required that the pole shall be on the curve, we must have, Art. (87),

$$b'^2 - 2pa' = 0,$$

and equation (1) reduces to

$$[r \sin^2 v + 2(b' \sin v - p \cos v)]r = 0,$$

which may be satisfied by placing $r = 0$, or

$$r \sin^2 v + 2(b' \sin v - p \cos v) = 0 \ldots\ldots\ldots\ldots(2).$$

Since the pole is on the curve, as at P, it is evident, that one value of r should be 0, whatever be the value of v; and that the other value, deduced from equation (2), should, as v is changed, give the distance of each point of the curve from the pole P.

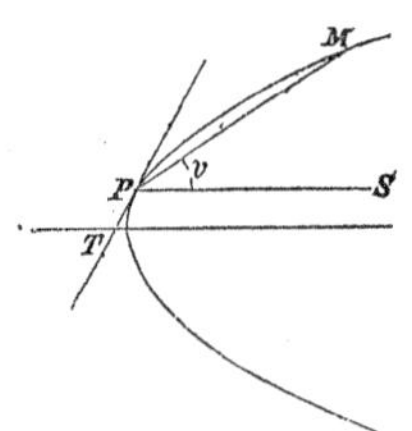

If the point M is moved along the curve until it coincides with P, the second value of r will become 0, and equation (2) will reduce to

$$b' \sin v - p \cos v = 0,$$

or

$$\frac{\sin v}{\cos v} = \text{tang } v = \frac{p}{b'},$$

as it should, Art. (90), since the radius vector will now coincide, in direction, with the tangent PT.

Second. If the pole be placed at the focus, we must have

$$a' = \frac{p}{2}, \qquad b' = 0,$$

and these values, in equation (1), give

$$r^2 \sin^2 v - 2p \cos vr - p^2 = 0,$$

and after transposing p^2 and dividing by $\sin^2 v$,

$$r^2 - \frac{2p \cos v}{\sin^2 v} r = \frac{p^2}{\sin^2 v}.$$

Solving this equation, we have

$$r = \frac{p \cos v}{\sin^2 v} \pm \sqrt{\frac{p^2}{\sin^2 v} + \frac{p^2 \cos^2 v}{\sin^4 v}},$$

or

$$r = \frac{p \cos v}{\sin^2 v} \pm \sqrt{\frac{p^2 (\sin^2 v + \cos^2 v)}{\sin^4 v}} = \frac{p \cos v \pm p}{\sin^2 v},$$

since $\sin^2 v + \cos^2 v = 1$.

As the $\cos v$ must be less than radius or unity, we have

$$p \cos v < p,$$

and the second value of r is always *negative*, and must therefore be rejected, Art. (69). The first value may be placed under the form

$$r = \frac{p (\cos v + 1)}{\sin^2 v},$$

and since

$$\sin^2 v = 1 - \cos^2 v = (1 + \cos v)(1 - \cos v),$$

it may be still further reduced to

$$r = \frac{p}{1 - \cos v} \ldots\ldots\ldots\ldots (3),$$

which is positive for all values of v.

If $v = 0$, $\cos v = 1$, and the value of r becomes

$$r = \frac{p}{0} = \infty,$$

and the radius vector takes the direction AX, and gives that point of the curve which is at an infinite distance.

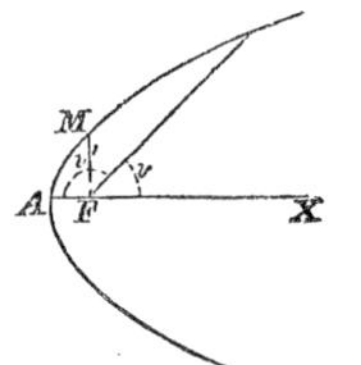

If $v = 90°$, $\cos v = 0$, and the value of r becomes

$$r = p = \text{FM}.$$

If $v = 180°$, $\cos v = -1$, and

$$r = \frac{p}{2} = \text{FA}.$$

Thus by varying v from 0 to 360°, all the points of the parabola may be determined.

If we wish to estimate the variable angle from the line FA, to the right, instead of in the usual way, from the line FX to the left, we have simply to change v into $180° - v'$, in which case $\cos v = -\cos v'$, and the value of r, equation (3), becomes

$$r = \frac{p}{1 + \cos v'},$$

in which $v' = 0$, gives $r = \frac{p}{2}$, and $v' = 180°$, gives $r = \infty$.

OF THE ELLIPSE AND HYPERBOLA.

105. We have seen, Art. (83), that the equation

$$y^2 = r^2x^2 + 2px, \qquad \text{or} \qquad y^2 = 2px + r^2x^2 \ldots\ldots(1),$$

represents the ellipse when r^2 is negative, and the hyperbola when it is positive.

This equation being of the second degree, the ellipse and hyperbola are both lines of the second order, Art. (33).

If in the equation, we make $x = 0$, we find

$$y = \pm 0;$$

hence the axis of Y is tangent to each curve, at the origin of co-ordinates, Art. (34).

If we make $y = 0$, we have

$$2px + r^2x^2 = 0, \qquad \text{or} \qquad x(2p + r^2x) = 0,$$

which may be satisfied by making

$$x = 0,$$

or

$$2p + r^2x = 0; \qquad \text{whence} \qquad x = -\frac{2p}{r^2};$$

hence each of the curves intersects the axis of X in two points, one at the origin, and the other at a distance from it equal to $-\frac{2p}{r^2}$.

Now let us transfer the origin of co-ordinates, to a point on the axis of X, at a distance $-\frac{p}{r^2}$, equal to half the distance from the origin to the second point in which the curve cuts the axis; the new axes being parallel to the primitive. In formulas (2), of Art. (67), we must then have

$$a' = -\frac{p}{r^2}, \qquad b' = 0,$$

and the formulas become

$$x = x' - \frac{p}{r^2}, \qquad y = y'.$$

Substituting these values of x and y in equation (1), we have

$$y'^2 = 2p\left(x' - \frac{p}{r^2}\right) + r^2\left(x'^2 - \frac{2px'}{r^2} + \frac{p^2}{r^4}\right),$$

or reducing and omitting the dashes,

$$y^2 = r^2x^2 - \frac{p^2}{r^2} \ldots\ldots\ldots\ldots(2).$$

If in this we make $y = 0$, we have

$$x^2 = \frac{p^2}{r^4} \ldots\ldots(3), \qquad \text{or} \qquad x = \pm\frac{p}{r^2};$$

hence, each curve now intersects the axis of X in two points, one on the right and the other on the left of the origin, at equal distances from it.

If $x = 0$, we have

$$y^2 = -\frac{p^2}{r^2} \ldots\ldots(4), \qquad \text{or} \qquad y = \pm\sqrt{-\frac{p^2}{r^2}},$$

and these values of y will be real for the ellipse, and imaginary for the hyperbola; hence, the ellipse intersects the axis of Y in two points, at equal distances from the origin, one above and the other below the axis of X; and the hyperbola does not intersect the axis of Y.

Giving to r^2 its negative sign for the ellipse, expressions (3) and (4) will be essentially positive, and we may write

$$\frac{p^2}{r^4} = a^2, \qquad - \frac{p^2}{r^2} = b^2;$$

from which, by deducing the values of p^2 and equating them, we have

$$a^2 r^4 = - r^2 b^2, \qquad \text{or} \qquad r^2 = - \frac{b^2}{a^2};$$

and substituting this in either of the above equations, we find

$$p^2 = \frac{b^4}{a^2}, \qquad \text{or} \qquad p = \frac{b^2}{a}.$$

By the substitution of these values of r^2 and p^2 in equation (2), and reducing, we have *the equation of the ellipse*,

$$a^2 y^2 + b^2 x^2 = a^2 b^2 \ldots\ldots\ldots\ldots (e).$$

For the hyperbola $- \dfrac{p^2}{r^2}$ is essentially negative, we must then place it equal to $- b^2$, while the expression for a^2 will remain unchanged. If then, in the above equation, we simply change b^2 into $- b^2$, we obtain *the equation of the hyperbola*,

$$a^2 y^2 - b^2 x^2 = - a^2 b^2 \ldots\ldots\ldots\ldots (h).$$

Furthermore, it is evident from the preceding discussion, that any expression containing b, belonging to the ellipse, will become the corresponding one for the hyperbola, by changing b^2 into $- b^2$, or b into $b\sqrt{-1}$.

106. Solving equation (e) with reference to y, we have

$$y^2 = \frac{b^2}{a^2}(a^2 - x^2), \qquad y = \pm \frac{b}{a}\sqrt{a^2 - x^2} \ldots\ldots (1),$$

in which every value of x numerically less than a, whether positive or negative, gives two real values of y equal with contrary signs: Hence, C being the origin, CX and CY the axes of co-ordinates, and CB and CA each numerically equal to a, the curve is continuous between the points A and B; and since each set of the equal values of y forms a chord as MM′, which is bisected by the axis of X, the curve is symmetrical with respect to the line AB.

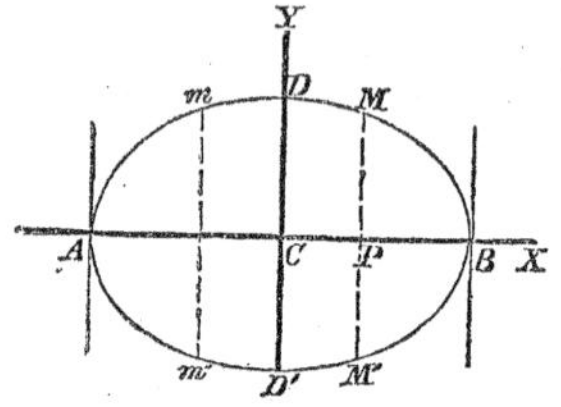

$x = + a$ or $- a$, gives

$$y = \pm 0;$$

hence the ordinates at the points A and B, when produced, are tangent to the curve. And as every value of x numerically greater than a, positive or negative, gives imaginary values for y, there are no points of the curve without the tangents at A and B.

$y = 0$ gives

$$x = \pm a = \text{CB or CA};$$

and since the line AB bisects a system of chords perpendicular to it, it is an axis of the curve, Art. (85), and A and B are its vertices.

$x = 0$ gives

$$y = \pm b = \text{CD or CD}'.$$

Any number of other points of the curve may be constructed by assigning values to x in equation (1), deducing and constructing the corresponding values of y, and the curve in form and position will be as in the last figure.

If equation (e) be solved with reference to x, we have

$$x = \pm \frac{a}{b} \sqrt{b^2 - y^2},$$

from which it may be shown as above, that the curve is symmetrical with respect to the axis of Y, and does not extend beyond the tangents at D and D′, and that the line DD′ is an axis of the curve.

The definite portion of the line AB, included within the ellipse, is called *the transverse axis*, and the portion DD′, *the conjugate axis;* the transverse axis being the longest of the two.

The point C, in which the axes intersect, is *the centre of the ellipse.*

The vertices of the transverse axis are also called *the vertices of the curve.*

Equation (e) is called *the equation of the ellipse referred to its centre and axes;* in which a represents the semi-transverse, and b the semi-conjugate axis.

107. If we solve equation (h), Art. (105), with reference to y, we have

$$y^2 = \frac{b^2}{a^2}(x^2 - a^2), \qquad y = \pm \frac{b}{a}\sqrt{x^2 - a^2},$$

in which every value of x numerically less than a, positive or negative, gives imaginary values of y: Hence, C being the origin, CX and CY the axes of coordinates, and CA and CB each numerically equal to a, there are no points of the curve between A and B.

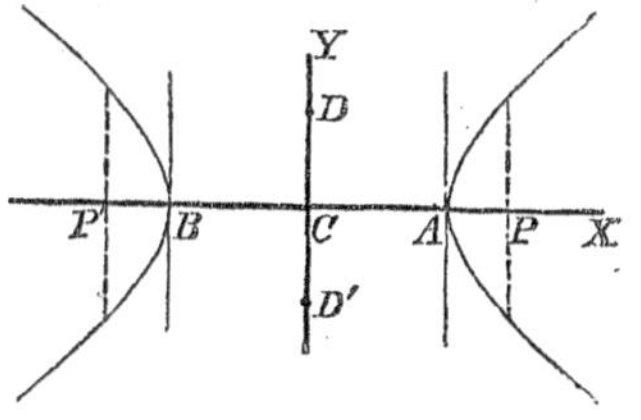

$x = +a$ or $-a$, gives

$$y = \pm 0;$$

hence, the ordinates at the points A and B, when produced, are tangent to the curve. Every value of x greater than a, positive or negative, gives two real values of y equal with contrary signs; hence, the curve is continuous and extends to infinity in both directions beyond the points A and B, and is symmetrical with respect to the axis of X.

$y = 0$ gives

$$x = \pm a = \text{CA or CB},$$

and since the line BA produced, bisects a system of chords perpendicular to it, it is an axis, and the definite portion $\text{BA} = 2a$, included between the points A and B, is called *the transverse axis* of the curve, the points A and B being its vertices or *the vertices of the curve.*

$x = 0$ gives

$$y^2 = -b^2, \qquad y = \pm b\sqrt{-1};$$

hence, the curve does not intersect the axis of Y.

A sufficient number of other points being constructed from the equation, the curve may be drawn as in the figure, the two branches being equal, since values of x which are numerically equal with contrary signs, as CP and CP′ give the same values for y.

If equation (h) be solved with reference to x, we have

$$x = \pm \frac{a}{b}\sqrt{y^2 + b^2},$$

in which *every value* of y gives two real values of x, equal with contrary signs; hence, the line CY is an axis of the curve. This line, as seen above, does not cut the hyperbola, but if we lay off on it from C, distances above and below each equal to b, the portion $\text{DD}' = 2b$ is called *the conjugate axis*, the point C being *the centre* of the hyperbola.

Equation (h) is called *the equation of the hyperbola referred to its centre and axes*, in which a and b represent the semi-axes.

108. If in equations (e) and (h), and, in general, in the equation of any curve, we change x into y and y into x, the effect is to change the line which at first is regarded as the axis of X, into the axis of Y and the converse; or if the axes are at right angles, to revolve the curve 90° about the origin. Thus if the equation

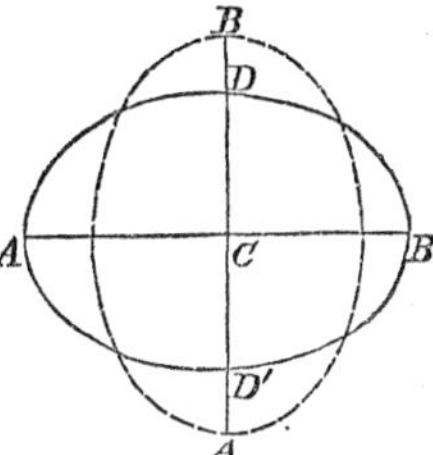

$$a^2y^2 + b^2x^2 = a^2b^2,$$

represents the ellipse as indicated by the full line, the equation

$$a^2x^2 + b^2y^2 = a^2b^2,$$

will represent it as indicated by the broken line.

109. If a point is on the ellipse, its co-ordinates must satisfy the equation of the ellipse, Art. (23), and we must have

$$a^2y^2 + b^2x^2 - a^2b^2 = 0.$$

If the point is without the ellipse, y will be greater than the corresponding ordinate of the ellipse, Art. (37), and we have

$$a^2y^2 + b^2x^2 - a^2b^2 > 0.$$

If it is within the ellipse

$$a^2y^2 + b^2x^2 - a^2b^2 < 0.$$

110. The corresponding conditions for the hyperbola, by changing, in the above, b^2 into $-b^2$, Art. (105), will be

$$a^2y^2 - b^2x^2 + a^2b^2 = 0.$$

$$a^2y^2 - b^2x^2 + a^2b^2 > 0.$$

$$a^2y^2 - b^2x^2 + a^2b^2 < 0.$$

111. If $a = b$, the axes of the ellipse are equal, and equation (e) becomes

$$y^2 + x^2 = a^2,$$

which is the equation of a circle, the radius of which is equal to either semi-axis, Art. (35).

112. Under the same supposition, equation (h) becomes

$$y^2 - x^2 = - a^2,$$

and the curve is called *an equilateral hyperbola.*

113. If through the centre of an ellipse any right line be drawn, its equation referred to the axes CX and CY, will be

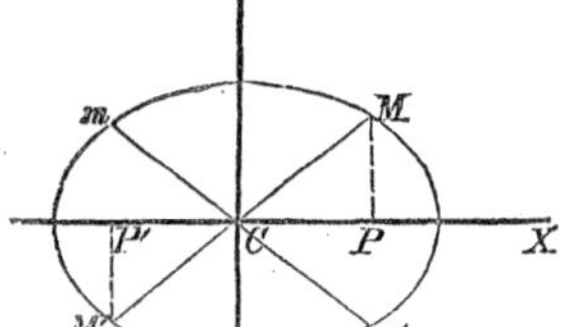

$$y = d'x \ldots\ldots\ldots\ldots (1,)$$

in which d' represents the tangent of the angle which the line makes with CX, Art. (24).

Combining this with equation (e), by substituting for y^2 its value d'^2x^2, we obtain

$$d'^2a^2x^2 + b^2x^2 = a^2b^2;$$

whence, for the abscissas of the points of intersection, Art. (27), we have

$$x = \pm \sqrt{\frac{a^2b^2}{d'^2a^2 + b^2}},$$

and by the substitution of this in equation (1),

$$y = \pm d' \sqrt{\frac{a^2b^2}{d'^2a^2 + b^2}}.$$

Since these values of x and y are real for every value of d', it follows that whatever be the position of the line CM, it will intersect the ellipse in two points; and since the co-ordinates of these points are equal with contrary signs, they will be on opposite sides of the origin, and at equal distances from it, as at M and M′. Hence, *every straight line passing through the centre of an ellipse and terminated by the curve is bisected at the centre.*

114. If in the above expressions we put $-b^2$ for b^2, the corresponding values for the hyperbola are

$$x = \pm\sqrt{\frac{-a^2b^2}{d'^2a^2 - b^2}}, \qquad y = \pm d'\sqrt{\frac{-a^2b^2}{d'^2a^2 - b^2}},$$

which are real, whenever

$$d'^2a^2 - b^2 < 0, \qquad \text{or} \qquad d' < \frac{b}{a},$$

that is, whenever d', either positive or negative, is numerically less than $\frac{b}{a}$, the line will cut the hyperbola in two points and be bisected at the centre. If

$$d' = \frac{b}{a},$$

the values are both infinite, and the points of intersection are at an infinite distance from C. If

$$d' > \frac{b}{a},$$

the values are imaginary, and the line will not intersect the curve. Hence, if at the point A, we erect the perpendiculars AE and AE′, each equal to b, and draw the lines CE and CE′, these lines will just limit the curve, since

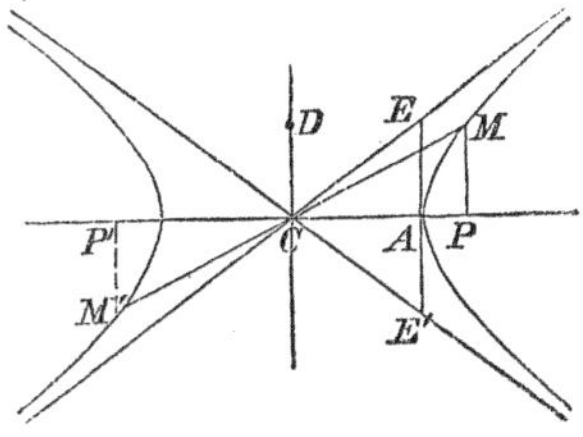

$$\text{tang ACE} = d' = \frac{\text{AE}}{\text{CA}} = \frac{b}{a}.$$

115. If we multiply both members of the expression $p = \frac{b^2}{a}$, Art. (105), by 2, we have

$$2p = \frac{2b^2}{a},$$

which as in the parabola, Art. (84), is called *the parameter*, and gives the proportion

$$a : b :: 2b : 2p, \qquad \text{or} \qquad 2a : 2b :: 2b : 2p.$$

Hence, *the parameter of the ellipse or hyperbola is a third proportional to the transverse and conjugate axes.*

116. If in equation (e), we substitute for y the expression $\frac{b^2}{a}$, we find

$$x^2 = a^2 - b^2, \qquad \text{or} \qquad x = \pm \sqrt{a^2 - b^2};$$

and conversely, if either of these values be substituted for x, we shall find

$$y = \pm \frac{b^2}{a};$$

from which we see, that there are two points on the transverse axis of the ellipse, at which, if an ordinate be drawn, it will be equal to one half the parameter of the curve; hence, *the double ordinate, or the chord perpendicular to the transverse axis, at each of these points, is equal to the parameter of the curve.*

These points are called *the foci of the ellipse*, and may be constructed thus: With either extremity of the conjugate axis as a centre, and the semi-transverse axis as a radius, describe an arc, the points in which this arc cuts the transverse axis will be the foci. For in the right angled triangle DCF or DCF′, we have, Art. (4),

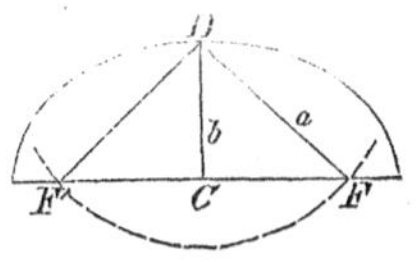

$$\overline{\text{CF}}^2 = \text{CF}'^2 = a^2 - b^2, \qquad \text{CF} = \text{CF}' = \pm\sqrt{a^2 - b^2}.$$

117. For the hyperbola, the values of x, in the preceding article, become

$$x = \pm\sqrt{a^2 + b^2}\text{............(1)},$$

either of which substituted in equation (h), will give

$$y = \pm\frac{b^2}{a},$$

and the points determined on the transverse axis, by laying off the above values of x are *the foci of the hyperbola*, and may be constructed thus: At either vertex of the hyperbola erect a perpendicular equal to b; join its extremity with the centre; with the last line CE, as a radius, and with the point C as a centre, describe an arc; the points in which this arc cuts the transverse axis produced, will be the foci. For we have

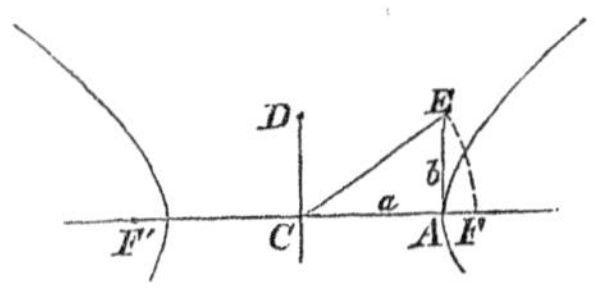

$$\overline{\text{CF}}^2 = \overline{\text{CE}}^2 = a^2 + b^2, \qquad \text{CF} = \text{CF}' = \pm\sqrt{a^2 + b^2}.$$

118. The distance from the centre to either focus of the ellipse,

divided by the semi-transverse axis, is called *the cccentricity of the curve*, the expression for which is

$$\frac{\sqrt{a^2 - b^2}}{a}.$$

If $a = b$, this reduces to 0; hence, the excentricity of a circle is nothing, and the foci are at the centre.

119. The expression for the eccentricity of an hyperbola, is

$$\frac{\sqrt{a^2 + b^2}}{a},$$

which, when $a = b$, becomes for the equilateral hyperbola, Art. (112),

$$\frac{a\sqrt{2}}{a} = \sqrt{2}.$$

120. If we denote the distance CF by c, and the distance from any point of the ellipse, as M, to the focus F, by r, the general expression for the square of this distance will be, Art. (17),

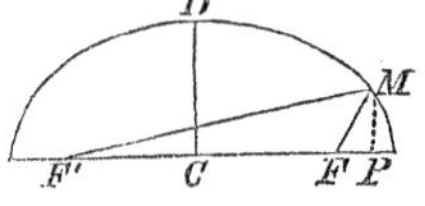

$$r^2 = (x - x')^2 + (y - y')^2,$$

in which

$$x' = c, \qquad y' = 0;$$

whence

$$\overline{FM}^2 = r^2 = (x - c)^2 + y^2,$$

and this, by the substitution of the values

$$y^2 = \frac{b^2}{a^2}(a^2 - x^2), \qquad c^2 = a^2 - b^2,$$

becomes

$$r^2 = x^2 - 2cx + a^2 - \frac{b^2}{a^2}x^2,$$

or

$$r^2 = \frac{a^4 - 2ca^2x + c^2x^2}{a^2},$$

and extracting the square root,

$$r = \frac{a^2 - cx}{a} = a - \frac{cx}{a} \ldots\ldots\ldots\ldots (1),$$

using the plus sign of the root only, as we require merely the expression for the length of FM.

Since $CF' = -c$; if in the above expression (1), we put $-c$ for c, we shall evidently obtain the distance F′M, which we denote by r'; hence,

$$r' = a + \frac{cx}{a} \ldots\ldots\ldots\ldots (2).$$

Adding equations (1) and (2), member by member, we have

$$r + r' = 2a;$$

hence, *the sum of the distances from any point of the curve to the two foci is equal to the transverse axis.*

This remarkable property enables us to define an ellipse to be, *a curve such, that the sum of the distances, from any point to two fixed points, is always equal to a given line.*

It also gives the following construction, of the curve by points, the foci and transverse axis being given. Divide the transverse axis into any two parts, the point of division being between the two foci, as at E; with one part EB

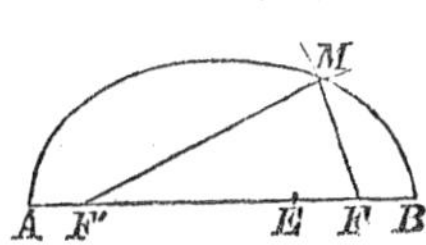

as a radius, and one focus F, as a centre, describe an arc; with the other part AE, as a radius, and the other focus as a centre, describe a second arc; the points of intersection of these arcs will be points of the ellipse. For we have

$$FM + F'M = EB + AE = 2a.$$

The curve may also be constructed by a continuous movement, thus: Take a thread, in length equal to the transverse axis, and fasten an end at each focus; press a pencil against the thread so as to draw it tight; the point of the pencil as it is moved around will describe the ellipse; for the sum of the distances, from this point to the foci, is always the same and equal to the transverse axis.

121. If F and F′ are the foci of the hyperbola, and the distances FM and F′M be denoted by r and r', we may deduce expressions for them from expressions (1) and (2) of the preceding article, by changing b^2 into $-b^2$, the only effect of which will be to make $c = \sqrt{a^2 + b^2}$ instead of $\sqrt{a^2 - b^2}$, and as for all points of the curve x must be greater than a, Art. (107), $\frac{cx}{a}$ must also be greater than a, and the expression for the numerical value of FM $= r$ will be

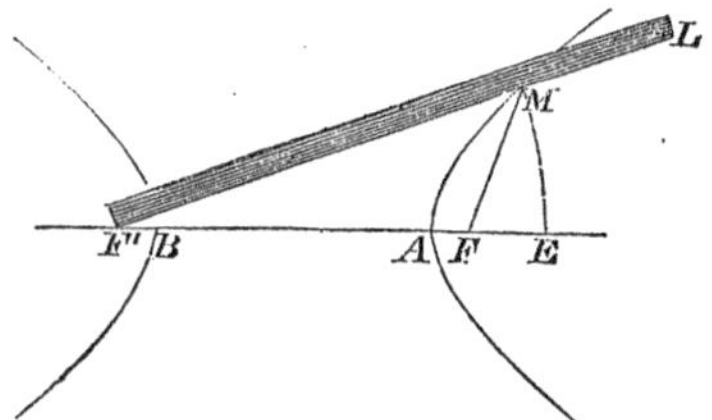

$$r = \frac{cx}{a} - a \ldots\ldots\ldots\ldots(3).$$

The form of the expression for r' will remain unchanged; hence,

$$r' = \frac{cx}{a} + a \ldots\ldots\ldots\ldots(4).$$

Subtracting the first of these from the second, we have

$$r' - r = 2a;$$

hence, *the difference of the distances from any point of the curve to the two foci, is equal to the transverse axis*; and the hyperbola may be defined to be, *a curve such, that the difference of the distances from any point to two fixed points is equal to a given line.*

The curve may be constructed by points thus: With one focus F′ as a centre, and any radius F′E, greater than the distance to the farther vertex, describe an arc; with the other focus and a radius FM, equal to the first radius minus the transverse axis, describe another arc; the points of intersection will be points of the curve. For, we have

$$F'M - FM = 2a.$$

It may also be constructed by a continuous movement. Take a rule of sufficient length as F′L, and fasten one end at the focus F′; at the other end of the rule fasten one end of a string shorter than the rule by the transverse axis; fasten the other end of the string at the other focus, F; press a pencil against the string and rule; as the rule revolves about the focus F′, the point of the pencil will describe the branch AM. For, we have

$$F'L - 2a = FM + ML,$$

or

$$F'L - ML - FM = 2a;$$

hence

$$F'M - FM = 2a.$$

By placing the end of the rule at F, the other branch may be described.

122. By a reference to equations (1) and (2), Art. (120), it is seen that *the distance from any point of the curve to either focus is expressed rationally in terms of its abscissa.*

This remarkable property of the foci is possessed by no other points in the plane of the curve. For, if there is any other point, let its co-ordinates be x' and y'; x and y denoting the co-ordinates of any point of the curve. The square of the distance from x, y, to x', y', Art. (17), is

$$D^2 = (x - x')^2 + (y - y')^2,$$

or squaring $x - x'$ and $y - y'$, and substituting for y its value

$$\pm \frac{b}{a} \sqrt{a^2 - x^2},$$

we have

$$D^2 = \frac{a^2 - b^2}{a^2} x^2 - 2xx' + x'^2 + b^2 \mp 2y' \frac{b}{a} \sqrt{a^2 - x^2} + y'^2.$$

It is evident that the value for D can not be rational, in terms of x, unless the term containing the radical disappears. But this can not be unless $y' = 0$, that is, the required point must be on the axis of X. Substituting this value for y', D^2, after changing the order of the terms, becomes

$$D^2 = (b^2 + x'^2) - 2xx' + \frac{a^2 - b^2}{a^2} x^2.$$

Now no value of x' can make this expression a perfect square unless it makes the first and last terms perfect squares, and twice the square root of their product equal to the middle term, that is, we must have

$$b^2 + x'^2 = m^2, \qquad \frac{a^2 - b^2}{a^2} x^2 = n^2, \qquad - 2xx' = 2mn,$$

m^2 and n^2 being two perfect squares. From the last expression we have

$$m^2 = \frac{x^2x'^2}{n^2},$$

in which, substituting the value of n^2 taken from the second, we have

$$m^2 = \frac{a^2x'^2}{a^2 - b^2}.$$

Substituting this in the first, we have

$$b^2 + x'^2 = \frac{a^2x'^2}{a^2 - b^2},$$

which can be satisfied for no values of x', except

$$x' = \pm \sqrt{a^2 - b^2},$$

the abscissas of the two foci.

In a similar way it may be shown, that the foci of the hyperbola and parabola alone possess the above named property.

123. If in equation (h) we substitute b for y, we deduce

$$x^2 = 2a^2, \qquad x = a\sqrt{2};$$

therefore, the abscissa of that point of the hyperbola, whose ordinate is equal to the semi-conjugate axis, is equal to the diagonal of a square, the side of which is the semi-transverse axis. Hence, the curve and transverse axis being given, the conjugate axis may be constructed thus: At the vertex A, erect a perpendicular AR $= a$; join the extremity

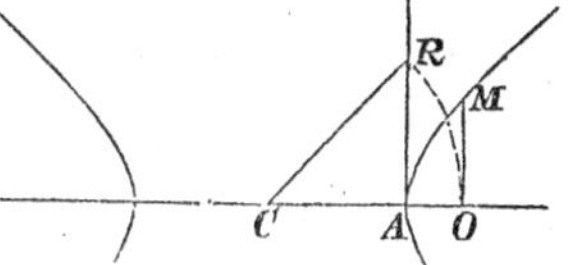

with the centre; with C as a centre, and CR as a radius, describe an arc cutting the transverse axis in O, at which erect the ordinate OM; it will be equal to the semi-conjugate axis. For we have

$$\text{CO} = \text{CR} = \text{CA}\sqrt{2} = a\sqrt{2}.$$

124. If the values

$$p = \frac{b^2}{a}, \qquad r^2 = \frac{b^2}{a^2},$$

Art. (105), be substituted in equation (1) of the same article, giving to r^2, first the negative and then the positive sign, we obtain the two equations

$$y^2 = \frac{b^2}{a^2}(2ax - x^2) \ldots\ldots\ldots\ldots (1),$$

$$y^2 = \frac{b^2}{a^2}(2ax + x^2) \ldots\ldots\ldots\ldots (2),$$

which are the equations of the ellipse and hyperbola referred to the axis and principal vertex A. See figures of Arts. (106), (107).

125. Let x', y', and x'', y'', be the co-ordinates of any two points of the ellipse. These co-ordinates, when substituted for x and y in equation (e), must satisfy it, Art. (23), and give the two equations of condition

$$a^2y'^2 + b^2x'^2 = a^2b^2, \qquad a^2y''^2 + b^2x''^2 = a^2b^2,$$

or

$$y'^2 = \frac{b^2}{a^2}(a^2 - x'^2) \ldots\ldots (1), \qquad y''^2 = \frac{b^2}{a^2}(a^2 - x''^2).$$

Dividing the first by the second, member by member, we have

$$\frac{y'^2}{y''^2} = \frac{a^2 - x'^2}{a^2 - x''^2} = \frac{(a + x')(a - x')}{(a + x'')(a - x'')};$$

whence, we deduce the proportion

$$y'^2 : y''^2 :: (a + x')(a - x') : (a + x'')(a - x'').$$

But

$$a + x' = \text{AP}, \qquad a - x' = \text{PB},$$

$$a + x'' = \text{AP}', \qquad a - x'' = \text{P}'\text{B}.$$

Therefore

$$\overline{\text{MP}}^2 : \overline{\text{M}'\text{P}'}^2 :: \text{AP} \times \text{PB} : \text{AP}' \times \text{P}'\text{B};$$

that is, *the squares of the ordinates of any two points of the ellipse are to each other as the rectangles of the segments into which they divide the transverse axis.*

For the circle, $a = b$, and equation (1) reduces to, Art. (36),

$$y'^2 = a^2 - x'^2 = (a + x')(a - x').$$

126. By using equation (h) and pursuing the same method as in the preceding article, we shall find for the hyperbola

$$y'^2 : y''^2 :: (x' + a)(x' - a) : (x'' + a)(x'' - a),$$

or

$$\overline{\text{MP}}^2 : \overline{\text{M}'\text{P}'}^2 : \text{AP} \times \text{BP} : \text{AP}' \times \text{BP}',$$

that is, *the squares of the ordinates of any two points of the hyperbola, are to each other as the rectangles of the distances from the foot of each ordinate to the vertices of the curve.*

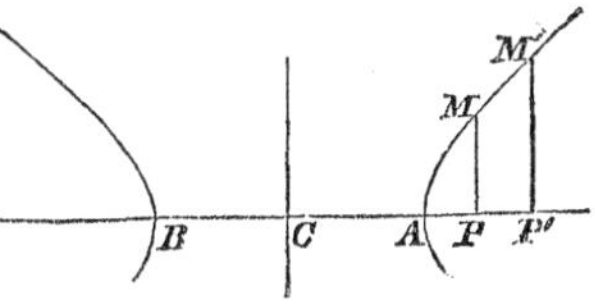

127. If with the centre of the ellipse as a centre, and $CA = a$ as a radius, a circle be described, its equation, Art. (111), may be put under the form

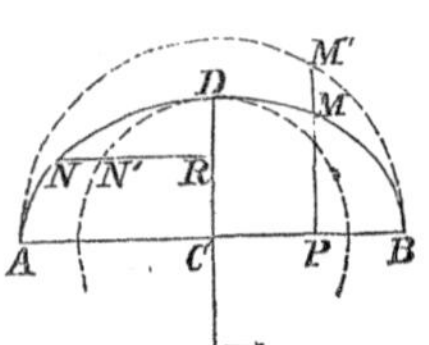

$$Y^2 = a^2 - x^2 \ldots\ldots\ldots\ldots(1),$$

in which Y represents the ordinate of any point of the circle, as M′P. From equation (e) we have

$$y^2 = \frac{b^2}{a^2}(a^2 - x^2) \ldots\ldots\ldots\ldots(2),$$

in which, if x have the same value as in equation (1), y will represent the ordinate MP, of the ellipse. Dividing equation (2) by (1), member by member, we have

$$\frac{y^2}{Y^2} = \frac{b^2}{a^2}, \qquad \frac{y}{Y} = \frac{b}{a};$$

whence

$$Y : y :: a : b,$$

that is, if a circle be described on the transverse axis of an ellipse, *any ordinate of the circle will be to the corresponding ordinate of the ellipse as the semi-transverse to the semi-conjugate axis.*

If with C as a centre, and $CD = b$ as a radius, a circle be described, its equation may be put under the form

$$X^2 = b^2 - y^2 \ldots\ldots\ldots\ldots (3),$$

in which X represents the abscissa of any point of the circle as RN′. If we obtain the value of x^2 from equation (e) and divide by equation (3), we may deduce the proportion

$$X : x :: b : a,$$

that is, if a circle be described on the conjugate axis of an ellipse, *any abscissa of the circle will be to the corresponding abscissa of the ellipse as the semi-conjugate to the semi-transverse axis.*

From the first of the above proportions, it appears that the ordinate of any point of the circle described on the transverse axis, is greater than the corresponding ordinate of the ellipse; hence, all the points of this circle are without the ellipse, except the vertices A and B.

From the second proportion, it also appears that every point of the circle described on the conjugate axis is within the ellipse, except the vertices D and D′.

We also conclude, that of all straight lines, passing through the centre, and terminating in the ellipse, the transverse axis is the longest, and the conjugate the shortest.

Upon the above properties, the following constructions of the ellipse depend.

First. On each of the axes as a diameter, describe a circle; at any point of the transverse axis, as P, erect a perpendicular and produce it, till it meets the outer circle in M′; join this point with the centre by the line M′C; from the point R, where this line meets the inner circle, draw a line parallel to the transverse axis, the point in which it meets the perpendicular will be a point of the ellipse. For, we have

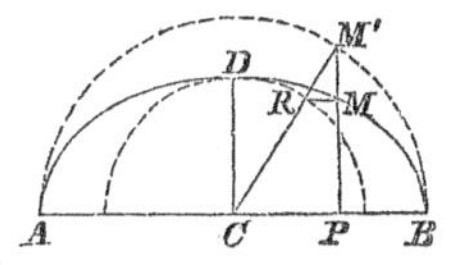

$$M'P : MP :: M'C : RC :: a : b.$$

Second. Take a rule MO, in length equal to the semi-transverse axis; from the extremity M, lay off MS equal to the semi-conjugate axis; move the rule so that the extremity O and the point of division S shall remain, the first on the conjugate and the second on the transverse axis; the point of a pencil at M, will describe the ellipse. For, draw OP′ parallel to CB,

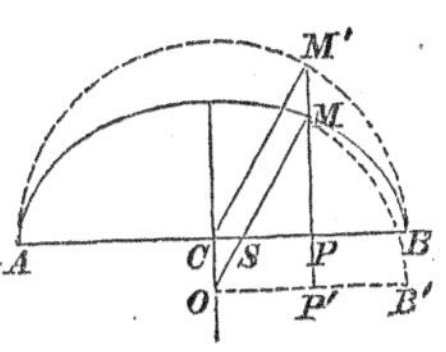

until it meets the produced ordinate MP in P′; join M′C, then the two equal right angled triangles M′CP and MOP′ give

$$MP' = M'P,$$

and the similar triangles MPS and MP′O give

$$MP' : MP :: MO : MS :: a : b.$$

128. Let x'', y'', be the co-ordinates of any point of the ellipse, as M, and through this point conceive any straight line to be drawn; its equation will be of the form

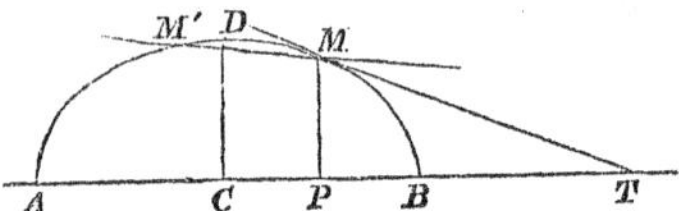

$$y - y'' = d(x - x'')...(1),$$

in which d is undetermined. Since the given point is on the curve, its co-ordinates must satisfy equation (e), and give the condition

$$a^2y''^2 + b^2x''^2 = a^2b^2.$$

Subtracting this, member by member, from equation (e), we have

$$a^2(y^2 - y''^2) + b^2(x^2 - x''^2) = 0,$$

or

$$a^2(y + y'')(y - y'') + b^2(x + x'')(x - x'') = 0.$$

Combining this with equation (1), by substituting the value of $y - y''$ taken from equation (1), we obtain

$$[da^2(y + y'') + b^2(x + x'')](x - x'') = 0,$$

in which x and y are the co-ordinates of all the points common to the right line and curve. This equation being of the second degree, there are two such points, and only two. These points may be determined by placing the factors, separately, equal to 0. Placing

$$x - x'' = 0, \qquad \text{we have} \qquad x = x'',$$

which in equation (1), gives

$$y = y'',$$

and these values evidently belong to the given point M. Placing the other factor equal to 0, we have

$$da^2(y + y'') + b^2(x + x'') = 0 \ldots\ldots\ldots\ldots (2),$$

in which x and y must be the co-ordinates of the second point of intersection M′.

If now the right line be revolved about the point M, until the point M′ coincides with M, the secant line will become a tangent; x and y, in equation (2), will become equal to x'' and y'', and the equation reduce to

$$2da^2y'' + 2b^2x'' = 0; \qquad \text{whence} \qquad d = -\frac{b^2x''}{a^2y''}.$$

Substituting this value of d in equation (1), we have

$$y - y'' = -\frac{b^2x''}{a^2y''}(x - x''),$$

which, since $a^2y''^2 + b^2x''^2 = a^2b^2$, reduces to

$$a^2yy'' + b^2xx'' = a^2b^2 \ldots\ldots\ldots\ldots (3),$$

for *the equation of a tangent line to the ellipse at a given point.*

If $a = b$, the above equation reduces to

$$yy'' + xx'' = a^2,$$

for the tangent line to the circle whose radius is a.

129. If we multiply both members of equation (3), preceding article, by 2, and subtract the result, member by member, from the equation

$$a^2y''^2 + b^2x''^2 = a^2b^2,$$

we have

$$a^2y''^2 - 2a^2yy'' + b^2x''^2 - 2b^2xx'' = - a^2b^2.$$

Adding $a^2y^2 + b^2x^2$, to both members, we have

$$a^2(y'' - y)^2 + b^2(x'' - x)^2 = a^2y^2 + b^2x^2 - a^2b^2.$$

The first member is the sum of two perfect squares, hence

$$a^2y^2 + b^2x^2 - a^2b^2$$

is positive for all values of x and y, except $x = x''$ and $y = y''$.

All points of the tangent, except the point of contact, are therefore without the ellipse, Art. (109).

130. If in equation (3), Art. (128), we make $y = 0$, we find

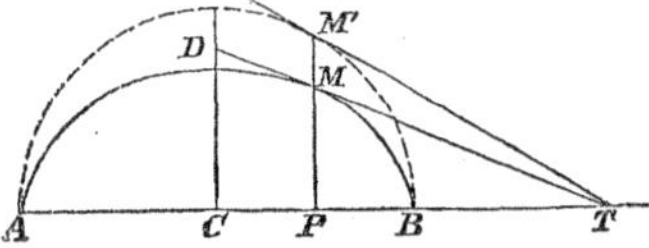

$$x = \frac{a^2}{x''} = \text{CT},$$

for the distance from C, to the point in which the tangent cuts the transverse axis. If from this we subtract the distance $\text{CP} = x''$, we have

$$\text{CT} - \text{CP} = \text{PT} = \frac{a^2}{x''} - x'' = \frac{a^2 - x''^2}{x''},$$

which is the subtangent, Art. (92). This expression for the subtangent, being independent of the conjugate axis, will be the same for all ellipses having the same transverse axis, and the points of contact in the same perpendicular to this axis. Hence, if it be required to draw a tangent to an ellipse at a given point as M: On the transverse axis describe a circle; through the given point draw a perpendicular to the axis and produce it until it meets the circle at M′; at this point draw a tangent to the circle, and connect the point T, in which this tangent cuts the axis produced, with the given point; this line will be the required tangent.

In a similar way, we may find the distance cut off by the tangent on the conjugate axis produced, and the expression for the subtangent on this axis.

131. If in equation (3), Art. (128), we change b^2 into $-b^2$, it becomes

$$a^2yy'' - b^2xx'' = -a^2b^2,$$

for *the equation of a tangent to the hyperbola at a given point.*

132. If in the last equation we make $y = 0$, we find

$$x = \frac{a^2}{x''} = \text{CT}......(1),$$

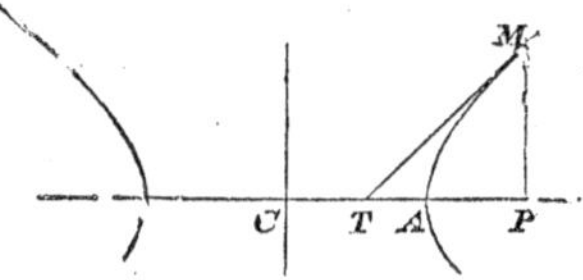

and subtracting this from $\text{CP} = x''$, we have

$$\text{CP} - \text{CT} = \text{PT} = \frac{x''^2 - a^2}{x''}$$

for the subtangent of the hyperbola.

133. Let MT be a tangent at any point M, of the ellipse, the co-ordinates of this point being x'' and y''; draw the lines MF and MF′ to the foci. In Art. (120), we have found

$$MF' = a + \frac{cx''}{a}, \qquad MF = a - \frac{cx''}{a},$$

or

$$MF' = \frac{a^2 + cx''}{a}, \qquad MF = \frac{a^2 - cx''}{a};$$

hence

$$MF' : MF :: a^2 + cx'' : a^2 - cx'' \ldots\ldots\ldots\ldots (1).$$

If to the expression $CT = \frac{a^2}{x''}$, Art. (130), we add $CF' = c$, and from it subtract $CF = c$, we have

$$F'T = \frac{a^2 + cx''}{x''}, \qquad FT = \frac{a^2 - cx''}{x''};$$

hence

$$F'T : FT :: a^2 + cx'' : a^2 - cx'',$$

and since the last terms of this proportion are the same as (1),

$$F'T : FT :: MF' : MF.$$

Through F draw FO parallel to MF', then

$$F'T : FT :: MF' : FO;$$

hence $FO = FM$, and the angle

$$FMO = FOM = F'MT'.$$

Therefore, *if from the point of contact of a tangent to an ellipse, two lines be drawn to the foci, these lines will make equal angles with the tangent.*

This property enables us to make the following constructions.

First. To draw a tangent to an ellipse at a given point. Join the point with the foci; produce the line F'M, drawn to one focus, until it is equal to the transverse axis; join its extremity B′, with the other focus; through the given point draw a line perpendicular to the last line; it will be the tangent. For,

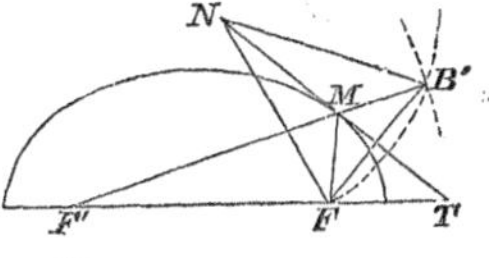

$$F'M + MB' = 2a = F'M + MF,$$

hence MB′ = MF, the triangle MFB′ is isosceles, and the angle

$$B'MT = F'MN = FMT.$$

Second. To draw a tangent to an ellipse from a point without the curve.

With either focus F′, as a centre, and radius equal to the transverse axis, describe an arc; with the given point N, as a centre, and radius equal to the distance to the other focus, describe another arc; join their point of intersection B′, with the first focus: the point M, in which this line intersects the ellipse, will be the point of contact, which being joined with the given point will give the tangent. For

$$NF = NB' \quad \text{and} \quad MF = MB';$$

hence the line NM, having two points at equal distances from F and B′, is perpendicular to FB′ at its middle point and bisects the angle FMB′. Since the two arcs above described intersect in two points, there will be two tangents.

Let the co-ordinates of the given point be x' and y'. Since it is on the tangent, we must have the condition, Art. (23),

$$a^2y'y'' + b^2x'x'' = a^2b^2 \ldots\ldots\ldots\ldots (2),$$

and since the point of contact is on the ellipse, we also have

$$a^2y''^2 + b^2x''^2 = a^2b^2 \ldots\ldots\ldots\ldots (3).$$

The combination of these equations will give two values of x

and two corresponding of y'', Art. (93), which will be of the form

$$y'' = m \pm n\sqrt{a^2y'^2 + b^2x'^2 - a^2b^2},$$

and these values will be real, equal, or imaginary as the given point is without, on, or within the ellipse, Art. (109), In the first case there will be two tangents, in the second but one, and in the third none.

134. Let MT be a tangent to the hyperbola at any point, and MF′ and MF lines drawn to the foci. In Art. (121), we have found

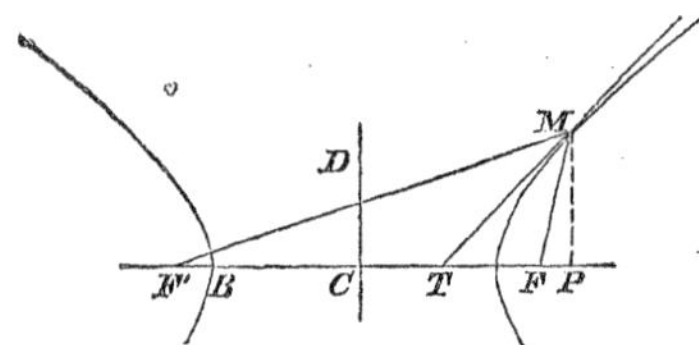

$$\text{MF}' = \frac{cx'' + a^2}{a},$$

$$\text{MF} = \frac{cx'' - a^2}{a}.$$

If to $c = \text{CF}' = \text{CF}$, we add the expression $\text{CT} = \frac{a^2}{x''}$, Art. (132), and then subtract it, we shall have

$$\text{F}'\text{T} = \frac{cx'' + a^2}{x''}, \qquad \text{FT} = \frac{cx'' - a^2}{x''}.$$

Hence, as in the preceding article, we deduce

$$\text{F}'\text{T} : \text{FT} :: \text{MF}' : \text{MF},$$

that is, the tangent MT divides the base of the triangle MF′F into two segments proportional to the adjacent sides, it therefore bisects the angle F′MF, at the vertex. Therefore, *if from the point of contact of a tangent to an hyperbola, two lines be drawn to the foci, these lines will make equal angles with the tangent.*

This property enables us to make the following constructions.

First. To draw a tangent to an hyperbola at a given point. Join the point M, with the foci; with the point as a centre and the distance to the nearest focus as a radius, describe an arc cutting the line drawn to the farthest focus in A′; join this point with the first focus and through the given point draw a line perpendicular to this last line; it will be the tangent. For the triangle MFA′ is isosceles; hence, the perpendicular MT bisects the angle F′MF.

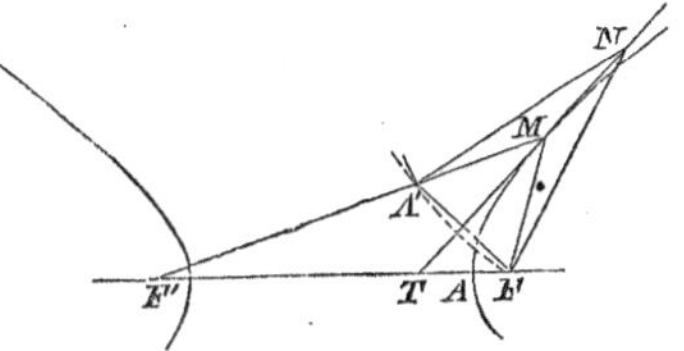

Second. To draw a tangent to an hyperbola from a point without the curve.

The construction and explanation of this are the same as for the ellipse.

If, as in the ellipse, the co-ordinates of the given point be denoted by x' and y', we shall have for the hyperbola the two equations of condition

$$a^2y'y'' - b^2x'x'' = -a^2b^2,$$

$$a^2y''^2 - b^2x''^2 = -a^2b^2,$$

the combination of which, will give two values of x'' and y'', which will be of the form

$$y'' = m \pm n\sqrt{a^2y'^2 - b^2x'^2 + a^2b^2},$$

and there will be two tangents, one, or none, as the given point is without, on, or within the hyperbola.

135. The general form of the equation of a straight line passing through the point B is, Art. (29),

$$y - y' = c\,(x - x'),$$

in which, for this particular case, we must have

$$y' = 0, \qquad x' = a,$$

which gives for the equation of BM,

$$y = c\,(x - a).$$

For the equation of the right line passing through A, for which

$$y' = 0, \qquad x' = \mp a,$$

we have

$$y = c'\,(x + a).$$

Combining these equations by multiplication, we have

$$y^2 = cc'\,(x^2 - a^2),$$

in which x and y are the co-ordinates of the point of intersection of the two lines, Art. (27). If this point is on the ellipse, x and y in the last equation must satisfy the equation of the ellipse, and we must also have

$$y^2 = \frac{b^2}{a^2}\,(a^2 - x^2) = -\frac{b^2}{a^2}\,(x^2 - a^2).$$

Equating these values of y^2, and omitting the common factor $x^2 - a^2$, we have

$$cc' = -\frac{b^2}{a^2} \ldots\ldots\ldots\ldots(1),$$

for *the equation of condition that the lines shall intersect on the ellipse.*

The lines when subjected to this condition are called *supplementary chords;* and, in general, *supplementary chords of a curve are straight lines drawn from the extremities of a diameter and intersecting on the curve.*

Since c and c' are indeterminate in the above equation of condition, an infinite number of supplementary chords can be drawn, and if any value be assigned to either c or c', the other becomes known and the position of the corresponding chord will be determined.

If $c = 0$, c' will be ∞; or if $c' = 0$, $c = \infty$; that is, if either chord coincides with the transverse axis, the other will be perpendicular to it.

If either c or c' is positive, the other must be negative; that is, if one chord makes an acute angle with the transverse axis, the other will make an obtuse, and the reverse.

If $a = b$, the condition (1) reduces to

$$cc' = -1, \qquad \text{or} \qquad cc' + 1 = 0;$$

hence, Art. (28), the supplementary chords of a circle are perpendicular to each other.

The expression for the tangent of the angle AMB is, Art. (28),

$$\text{tang V} = \frac{c - c'}{1 + cc'}.$$

But since c is the tangent of the obtuse angle MBX, it is essentially negative and may be placed $= -c''$. Substituting this, and also $cc' = -\frac{b^2}{a^2}$, the above expression becomes

$$\text{tang V} = \frac{-(c'' + c')}{1 - \frac{b^2}{a^2}},$$

which is essentially negative for all values of c and c'; hence the supplementary chords, drawn from the extremities of the transverse axis of an ellipse, make an obtuse angle with each other.

As the angle V is obtuse, it will be the greatest when its tangent is numerically the least; and since the denominator of the above expression is constant, it will be the least when the numerator is the least. But the product of c' and c'' being constant, their sum $c'' + c'$, will be the least when the factors are equal,* that is, when

$$c'' = c', \qquad \text{or} \qquad -c = c',$$

in which case, the angles are supplements of each other and the chords are drawn to the extremity of the conjugate axis D, making the angle DAC = DBC.

136. If we put $-b^2$ for b^2 in condition (1) of the preceding article, we obtain

$$cc' = \frac{b^2}{a^2},$$

for the equation of condition for supplementary chords drawn from the extremities of the transverse axis of the hyperbola.

As in the ellipse, an infinite number of chords may be drawn, and if either c or c' is positive or negative, the other must have the same sign; that is, both angles are at the same time acute, or both obtuse.

If $a = b$, the above equation becomes

---

* NOTE.—To prove this, let s represent their sum and d their difference, then

$$\frac{s}{2} + \frac{d}{2} = \text{the greater}, \qquad \frac{s}{2} - \frac{d}{2} = \text{the less},$$

and

$$\frac{s^2}{4} - \frac{d^2}{4} = \text{the product} = \text{P},$$

or

$$\frac{s^2}{4} = \text{P} + \frac{d^2}{4};$$

from which we see that s^2 or s will be the least when $d = 0$, or the two factors are equal.

$$cc' = 1, \qquad \text{or} \qquad c = \frac{1}{c'};$$

hence, in the equilateral hyperbola the two angles are complements of each other.

The expression for the tangent of the angle BMA, is

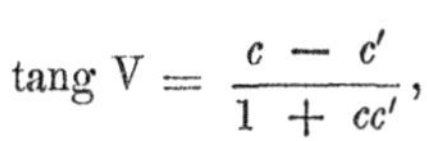

$$\text{tang V} = \frac{c - c'}{1 + cc'},$$

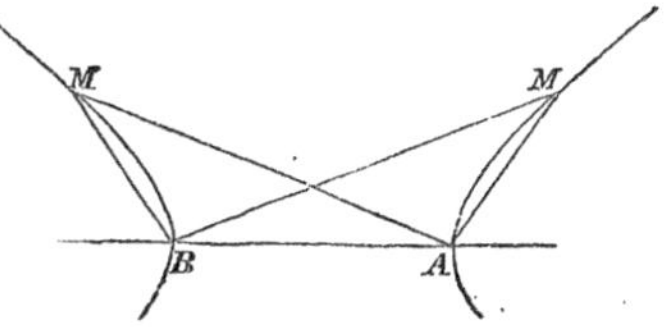

which is essentially positive, since c is always greater than c'; hence, the supplementary chords make an acute angle with each other, and this angle increases as c increases, until the chord AM becomes perpendicular to the transverse axis at the vertex A, when the angle is the greatest possible and equal to 90°.

137. If a right line be drawn through the centre of the ellipse, its equation will be

$$y = d'x,$$

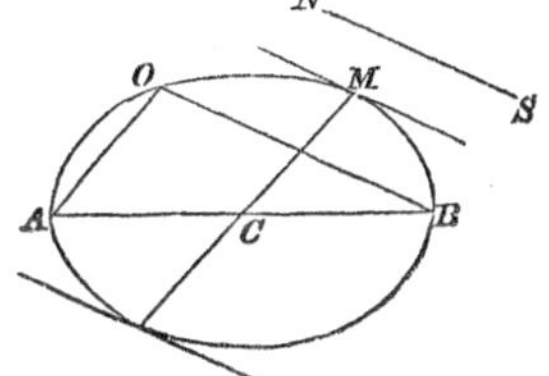

and if it pass through the point of contact of a tangent, we shall have the condition

$$y'' = d'x'', \qquad \text{or} \qquad d' = \frac{y''}{x''}.$$

Multiplying this, member by member, by the expression for d, Art. (128,) we have

$$dd' = -\frac{b^2}{a^2}\text{............}(1),$$

the same expression as that found in Art. (135) for cc'; hence

$$cc' = dd',$$

in which, if $c = d$, c' will be equal to d', and if $c' = d'$, $c = d$.

Therefore, *if one of the supplementary chords of an ellipse is parallel to a tangent, the other will be parallel to a line joining the point of contact and the centre, and the converse.*

Upon this property the following constructions depend.

First. To draw a tangent to an ellipse at a given point. From the point, draw a line, MC, to the centre; from one extremity of the transverse axis draw a chord, AO, parallel to this line; draw the supplement BO, of this chord, and at the given point, draw a line parallel to this supplement, it will be the required tangent.

Second. To draw a tangent to the ellipse parallel to a given line.

From one extremity of the transverse axis, draw a chord, BO, parallel to the given line NS; draw the supplement of this chord AO; parallel to which draw a line, CM, through the centre; at the points in which this line intersects the curve, draw lines parallel to the given line, they will be the required tangents.

138. By changing b^2 into $-b^2$, in the expression for d, Art. (128), it becomes the tangent of the angle made with the transverse axis by a tangent to the hyperbola, and by using this expression with the equation of condition

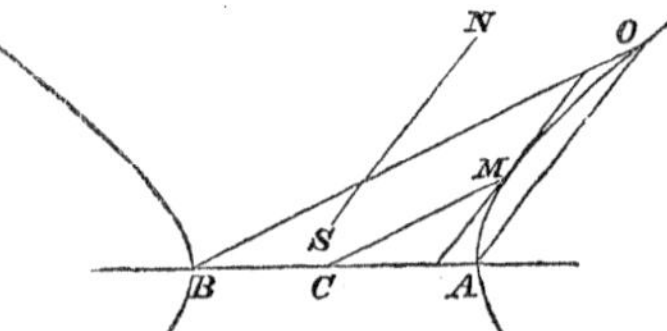

$$y'' = d'x'',$$

we have a similar discussion, and deduce the same properties of supplementary chords, and the same constructions for tangent lines as in the ellipse, as indicated in the figure.

It will evidently be impossible to draw a tangent to the hyperbola parallel to a given line, when the diameter to be drawn parallel to the second chord, does not intersect the curve.

139. If in the equation

$$a^2y'y'' + b^2x'x'' = a^2b^2 \ldots\ldots\ldots\ldots(1),$$

Art. (133), x'' and y'' be regarded as variables, it will be the equation of a right line; and since both values of x'' and y'' deduced from equations (2) and (3), Art. (133), must satisfy this equation, the right line must pass through both points of contact, or will be the indefinite chord which joins them.

If any point, as O, be taken upon this chord, its co-ordinates, which we denote by c and d, will satisfy equation (1), and give the condition

$$a^2y'd + b^2x'c = a^2b^2 \ldots\ldots\ldots\ldots(2).$$

Every set of values for x' and y' which will satisfy this equation, will give a point from which, if two tangents be drawn to the ellipse, the chords joining the points of contact will pass through the point O. Hence, if y' and x' be regarded as variables in this equation, it will represent a right line, every point of which will fulfil the above condition.

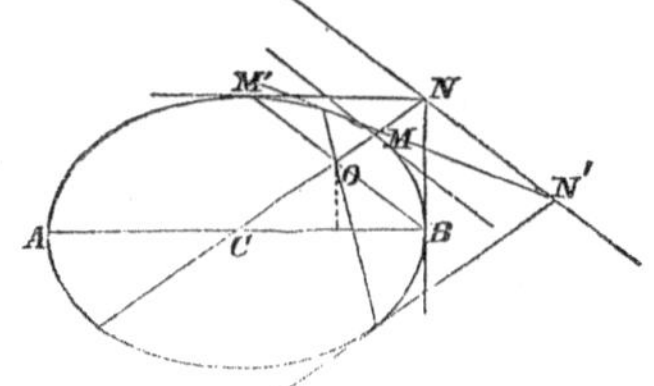

As in Art. (95), this line is the polar line of the pole O.

If through the point O and the centre, a right line be drawn, its equation will be

$$y = \frac{d}{c}x.$$

If this equation be combined with the equation of the ellipse, (e), Art. (105), we find for the co-ordinates of the point M,

$$x = \frac{abc}{\sqrt{a^2d^2 + b^2c^2}}, \qquad y = \frac{abd}{\sqrt{a^2d^2 + b^2c^2}}.$$

Substituting these for x'' and y'' in the equation of the tangent line, (3), Art. (128), we have for the equation of the tangent at the point M,

$$a^2dy + b^2cx = ab\sqrt{a^2d^2 + b^2c^2},$$

which is evidently parallel to the polar line, represented by equation (2).

If the line OC be produced until it intersects the polar line NN′ in N; for this point we shall have

$$\frac{x'}{y'} = \frac{c}{d} \quad \text{and} \quad \frac{b^2x'}{a^2y'} = \frac{b^2c}{a^2d};$$

hence, the chord which joins the points of contact, M′ and B, of two tangents drawn from N, in this case represented by equation (1), will also be parallel to the polar line.

These properties give the following constructions.

First. The pole being given, to construct the corresponding polar line.

Through the pole and centre, draw the line OC; at the point M, in which it intersects the curve, draw a tangent; through the pole draw a chord parallel to this tangent; at either point, as M′, in which this chord intersects the curve, draw a second tangent; through the point N, in which this intersects the line CO produced, draw a line parallel to the first tangent, it will be the required polar line.

Second. The polar line being given, to construct the corresponding pole.

Draw a tangent parallel to the polar line; join the point of contact M, with the centre, and produce this line until it meets the polar line in N; through this point draw a second tangent NM′, and through the point of contact, M′, draw a chord M′O, parallel to the polar line; the point in which it intersects the line MC will be the pole.

It should be remarked, that if the given line cuts the ellipse, this construction will fail, as the point N will lie within the ellipse and no tangent can be drawn from it.

When the ellipse becomes a circle, the line CM becomes perpendicular to the tangent at M and also to the polar line, and the above constructions are much simplified.

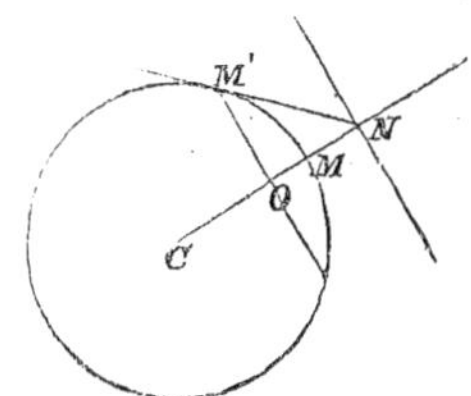

Thus, to construct the polar line: Through the given pole draw a line to the centre; draw a second line perpendicular to this, at the pole; at either point in which this perpendicular intersects the circle draw a tangent; through the point N, in which this tangent intersects the line drawn to the centre, draw a line perpendicular to the last line; it will be the polar line.

To construct the pole: Through the centre draw a line perpendicular to the polar line; from the point in which it intersects it, draw a tangent; from the point of contact draw a perpendicular to the first line; the point in which it intersects it will be the pole.

140. The equations of the preceding article become the corresponding equations of the hyperbola, by changing b^2 into $-b^2$, and it will be readily seen that the properties of the polar line and the constructions are precisely the same as for the ellipse.

When it is impossible to draw a tangent to the hyperbola parallel to a given line, Art. (138), the construction will fail.

141. The equation of any straight line passing through the point of contact of a tangent to an ellipse, will be of the form

$$y - y'' = d'(x - x'')\text{............}(1).$$

If this line is perpendicular to the tangent, we must have, Art. (28),

$$dd' + 1 = 0, \qquad \text{or} \qquad d' = -\frac{1}{d}.$$

But, Art. (128),

$$d = -\frac{b^2x''}{a^2y''};$$

whence

$$d' = \frac{a^2y''}{b^2x''},$$

and equation (1) becomes

$$y - y'' = \frac{a^2y''}{b^2x''}(x - x'')\text{............}(2),$$

for *the equation of a normal to the ellipse*, Art. (98).

If we make $y = 0$, in equation (2), we deduce

$$x'' - x = \frac{b^2x''}{a^2},$$

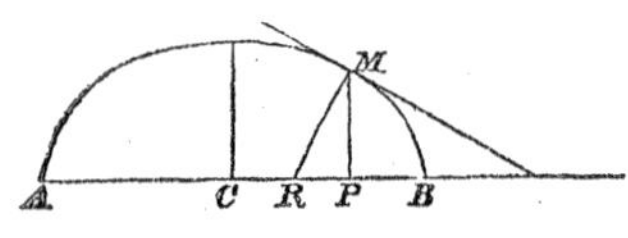

in which x is the distance CR, and

$$x' - x = \text{CP} - \text{CR} = \text{RP} = \textit{the subnormal}.$$

If $a = b$, equation (2) becomes

$$y - y'' = \frac{y''}{x''}(x - x'')$$

or

$$x''y - y''x = 0.$$

As there is no absolute term to this equation, the normal to the circle passes through the centre, Art. (38).

142. On the transverse axis of the ellipse let a semi-circle be described, and within this semi-circle let us inscribe any polygon, AN′NB. From the vertices of this polygon draw ordinates to the transverse axis, and join the points in which they intersect the ellipse, thus forming a polygon AM′MB, of the same number of sides.

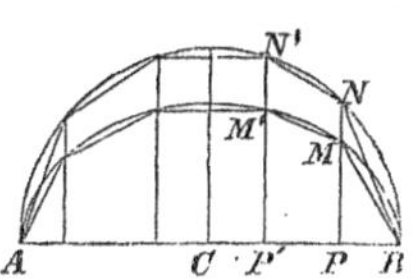

If the ordinates of the points N, N′, &c., be denoted by Y, Y′, &c., and the corresponding ordinates of M, M′, by y, y', &c., the abscissas being x, x', &c., we shall have, Art. (127),

$$\mathrm{Y} : y :: a : b, \qquad \mathrm{Y}' : y' :: a : b;$$

whence

$$\mathrm{Y} + \mathrm{Y}' : y + y' :: a : b.$$

The area of the trapezoid PNN′P′, forming a part of the polygon in the circle, will be

$$\left(\frac{\mathrm{Y} + \mathrm{Y}'}{2}\right)(x - x'),$$

and the area of the corresponding trapezoid, PMM′P′,

$$\left(\frac{y + y'}{2}\right)(x - x').$$

These expressions being equi-multiples of $\mathrm{Y} + \mathrm{Y}'$ and $y + y'$, are to each other as

$$\mathrm{Y} + \mathrm{Y}' : y + y', \qquad \text{or as} \qquad a : b.$$

In the same way, it may be proved that any trapezoid in the

circle is to the corresponding one in the ellipse as a is to b; hence, the sum of all in the circle, or the polygon, AN'NB, will be to the sum of all in the ellipse, or the corresponding polygon AM'MB, as a is to b; and this will be true, whatever be the number of the sides.

If now the number of sides be indefinitely increased, the areas of the polygons will become equal to the areas of the circle and ellipse respectively, and we shall have the first is to the second as a is to b; or denoting the area of the circle by S, and that of the ellipse by s, we shall have

$$S : s :: a : b; \qquad \text{whence} \qquad s = \frac{b}{a} S,$$

and substituting for S its value πa^2,

$$s = \pi ab,$$

or *the area of an ellipse is equal to the rectangle upon its semi-axes multiplied by the ratio of the diameter to the circumference of a circle.*

The above expression may be put under the form

$$s = \pi ab = \sqrt{\pi^2 a^2 b^2} = \sqrt{\pi a^2 \times \pi b^2},$$

that is, the area of the ellipse is a mean proportional between the areas of the two circles described, one upon the transverse, and the other upon the conjugate axis.

OF CONJUGATE DIAMETERS OF THE ELLIPSE AND HYPERBOLA.

143. Let it now be proposed to ascertain if there are any other co-ordinate axes, having their origin at the centre, to which, if the ellipse and hyperbola be referred, their equations will have the same form as when referred to their centres and axes, Arts. (106)

and (107). For this purpose let us take formulas (3), Art. (67), and substitute the values of x and y in equation (e), we thus obtain

$$a^2(x'^2 \sin^2 \alpha + 2x'y' \sin \alpha \sin \alpha' + y'^2 \sin^2 \alpha')$$

$$+ b^2(x'^2 \cos^2 \alpha + 2x'y' \cos \alpha \cos \alpha' + y'^2 \cos^2 \alpha') = a^2b^2,$$

or arranging and omitting the dashes of the variables,

$$(a^2 \sin^2 \alpha' + b^2 \cos^2 \alpha')y^2 + (a^2 \sin^2 \alpha + b^2 \cos^2 \alpha)x^2$$

$$+ 2(a^2 \sin \alpha \sin \alpha' + b^2 \cos \alpha \cos \alpha')xy = a^2b^2 \ldots\ldots\ldots\ldots(1),$$

which is the equation of the ellipse referred to any set of oblique axes, having the origin at the centre. This equation will be of the same form as equation (e), if the term containing xy be made to disappear, which requires that

$$a^2 \sin \alpha \sin \alpha' + b^2 \cos \alpha \cos \alpha = 0 \ldots\ldots\ldots\ldots(2).$$

The substitution of this condition in equation (1), reduces it to

$$(a^2 \sin^2 \alpha' + b^2 \cos^2 \alpha')y^2 + (a^2 \sin^2 \alpha + b^2 \cos^2 \alpha)x^2 = a^2b^2 \ldots(3).$$

Making y and x, in succession, each equal to 0, we find

$$x = \pm \sqrt{\frac{a^2b^2}{a^2 \sin^2 \alpha + b^2 \cos^2 \alpha}} = \text{CB}',$$

$$y = \pm \sqrt{\frac{a^2b^2}{a^2 \sin^2 \alpha' + b^2 \cos^2 \alpha'}} = \text{CD}',$$

both of which values are real for all values of α and α'; hence, the curve cuts each axis of co-ordinates in two points, on different sides of the centre, and at equal distances.

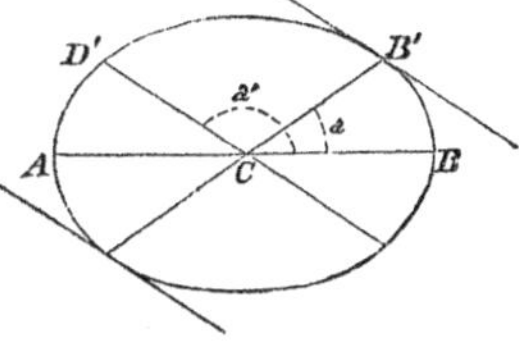

If we place these distances respectively equal to a' and b', we have

$$a'^2 = \frac{a^2b^2}{a^2 \sin^2 \alpha + b^2 \cos^2 \alpha}, \quad b'^2 = \frac{a^2b^2}{a^2 \sin^2 \alpha' + b^2 \cos^2 \alpha'}$$

from which

$$a^2 \sin^2 \alpha + b^2 \cos^2 \alpha = \frac{a^2b^2}{a'^2}, \quad a^2 \sin^2 \alpha' + b^2 \cos^2 \alpha' = \frac{a^2b^2}{b'^2}.$$

Substituting these values for the coefficients of x^2 and y^2 in equation (3), and striking out the common factor a^2b^2, we have

$$\frac{y^2}{b'^2} + \frac{x^2}{a'^2} = 1, \quad \text{or} \quad a'^2y^2 + b'^2x^2 = a'^2b'^2 \ldots\ldots (e'),$$

an equation of precisely the same form as equation (e), and which if solved as in Art. (106), will give for each value of $x < a'$, two values of y equal with contrary signs, and these taken together will form a chord mm', which is bisected by the axis of X; hence, this axis is a diameter of the ellipse, Art. (100). By solving equation (e') with reference to x, it may also be proved that the axis of Y is a diameter and bisects a system of chords parallel to the axis of X. These diameters are called conjugate diameters; and in general, *two diameters are conjugate, when each bisects a system of chords parallel to the other.*

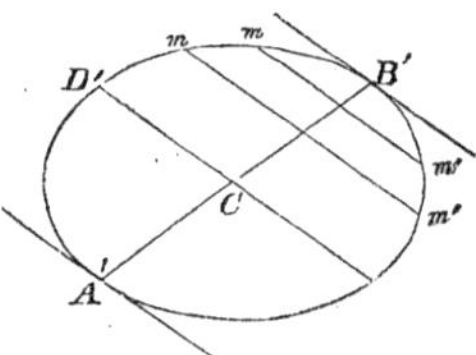

If in equation (e') we make $x = \pm a'$, we deduce

$$y = \pm 0;$$

hence, the ordinates at A′ and B′, produced, are tangent to the curve, Art. (34).

If $y = \pm b'$,

$$x = \pm 0.$$

Hence, the tangent, at the vertex of either diameter, is parallel to its conjugate, or, *to the chords which the diameter bisects.*

Equation (e') is called the equation of the ellipse referred to its centre and conjugate diameters, in which a' and b' are the semi-conjugate diameters.

144. Since, whenever α and α' have such values as to satisfy equation (2), of the preceding article, the axes of co-ordinates become conjugate diameters, that equation is called, *the equation of condition for conjugate diameters,* in which α and α' are the angles formed by these diameters respectively, with the transverse axis.

Dividing by $\cos \alpha \cos \alpha'$, and recollecting that

$$\frac{\sin \alpha}{\cos \alpha} = \text{tang } \alpha, \qquad \frac{\sin \alpha'}{\cos \alpha'} = \text{tang } \alpha',$$

we may put the equation under the form

$$\text{tang } \alpha \text{ tang } \alpha' = -\frac{b^2}{a^2} \ldots\ldots\ldots\ldots (1).$$

Since α and α' are indeterminate in this equation, it follows that there is an infinite number of conjugate diameters, and if a particular value be assigned to α or α' the corresponding value of the other will be determined and the position of the diameters known.

If $\alpha = 0$, $\text{tang } \alpha = 0$, and equation (1) gives

$$\text{tang } \alpha' = \infty, \qquad \alpha' = 90°.$$

If $\alpha' = 0$, $\text{tang } \alpha' = 0$, whence

$$\text{tang } \alpha = \infty, \qquad \alpha = 90°.$$

Hence, if either diameter coincides with the transverse axis the other wi l coincide with the conjugate. Also, if either α or α' is 90° the other will be 0; that is, if either diameter coincides with the conjugate axis, the other will coincide with the transverse; and the axes are conjugate diameters.

145. If any conjugate diameters, except the axes, are at right angles, we must have, [see figure of Art. (143)],

$$\text{D}'\text{CB}' = \alpha' - \alpha = 90^\circ, \quad \alpha' = 90^\circ + \alpha, \quad \text{tang}\, \alpha' = -\cot \alpha.$$

By the substitution of the last value in equation (1), of the preceding article, it becomes

$$\text{tang}\, \alpha \cot \alpha = \frac{b^2}{a^2} \text{.............}(1),$$

which, (since from Trigonometry $\text{tang}\, \alpha \cot \alpha = \text{R}^2 = 1$) can be satisfied for no value of α, except $\alpha = 0$, or $\alpha = 90^\circ$, in which case as seen above, the diameters coincide with the axes; hence, *the axes are the only conjugate diameters at right angles.*

If $a = b$, equation (1) becomes

$$\text{tang}\, \alpha \cot \alpha = 1,$$

which is satisfied for any value of α; hence in a circle, any two conjugate diameters are at right angles.

146. By comparing equation (1), Art. (144), with equation (1), Art. (135), we see that

$$cc' = \text{tang}\, \alpha \ \text{tang}\, \alpha';$$

hence, if $c = \text{tang}\, \alpha$,

$$c' = \text{tang}\, \alpha',$$

and the reverse; that is, *if one of two supplementary chords is parallel to a diameter, the other will be parallel to its conjugate.*

147. If in equation (1) of Art. (144), we put $-b^2$ for b^2, we have

$$\tang \alpha \ \tang \alpha' = \frac{b^2}{a^2} \ldots\ldots\ldots\ldots(1),$$

which is *the equation of condition for conjugate diameters in the hyperbola*, and admits of the same discussion, and gives precisely the same results for the hyperbola, as were deduced above for the ellipse.

If $a = b$, we have

$$\tang \alpha = \frac{1}{\tang \alpha'} = \cot \alpha';$$

hence, in the equilateral hyperbola, the conjugate diameters form angles with the transverse axis, which are complements of each other.

148. If in equation (3), Art. (143), we put $-b^2$ for b^2, it becomes

$$(a^2 \sin^2 \alpha' - b^2 \cos^2 \alpha')y^2 + (a^2 \sin^2 \alpha - b^2 \cos^2 \alpha)x^2 = -a^2b^2 \ldots(1),$$

and making y and x, in succession, each equal to 0, we find

$$x = \pm \sqrt{\frac{-a^2b^2}{a^2 \sin^2 \alpha - b^2 \cos^2 \alpha}},$$

$$y = \pm \sqrt{\frac{-a^2b^2}{a^2 \sin^2 \alpha' - b^2 \cos^2 \alpha'}}.$$

The reality of these values will depend upon the sign of the denominator under the radical sign. If that of the first is negative, x will be real. In this case

$$a^2 \sin^2 \alpha - b^2 \cos^2 \alpha < 0, \qquad \frac{\sin^2 \alpha}{\cos^2 \alpha} < \frac{b^2}{a^2}, \qquad \tang \alpha < \frac{b}{a};$$

hence, from equation (1) of the preceding article, we have

$$\text{tang}\ \alpha' > \frac{b}{a}, \quad \frac{\sin^2 \alpha'}{\cos^2 \alpha'} > \frac{b^2}{a^2}, \quad a^2 \sin^2 \alpha' - b^2 \cos^2 \alpha' > 0,$$

and the denominator, under the second radical sign, is positive, and the value of y imaginary.

In the same way, it may be shown that if y is real, x must be imaginary. Therefore, if one of the conjugate diameters of the hyperbola cuts the curve the other will not, and the converse.

If then, we regard the above value of x as real, we may place it equal to a', and the imaginary value of y equal to $b'\sqrt{-1}$, whence

$$a'^2 = \frac{-a^2b^2}{a^2 \sin^3 \alpha - b^2 \cos^2 \alpha}, \quad -b'^2 = \frac{-a^2b^2}{a^2 \sin^2 \alpha' - b^2 \cos^2 \alpha'},$$

from which, deducing the values of the denominators, and substituting in equation (1), we have

$$\frac{y^2}{b'^2} - \frac{x^2}{a'^2} = -1, \quad \text{or} \quad a'^2y^2 - b'^2x^2 = -a'^2b'^2 \ldots\ldots (h'),$$

for the equation of the hyperbola referred to its centre and conjugate diameters, in which, a' and b' are the semi-conjugate diameters.

This equation is of the same form as equation (h), Art. (107), and from it we may prove as in Art. (143), that each diameter bisects a system of chords parallel to its conjugate, or parallel to the tangent at its vertices, if it have vertices. If a second hyperbola be described upon DE as a transverse axis, having BA for its conjugate, it is said to be conjugate to the first

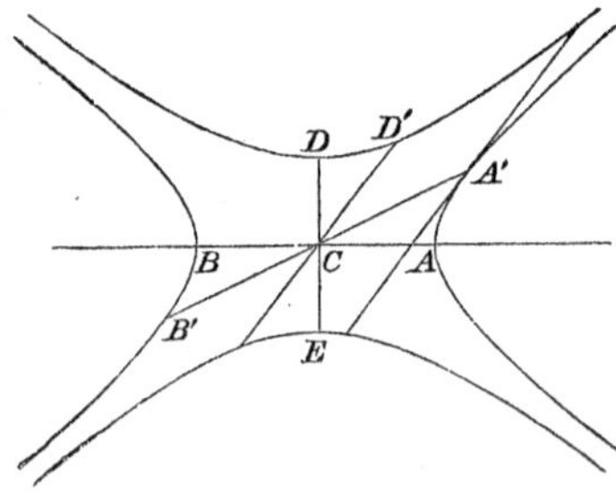

hyperbola; that is, *two hyperbolas are conjugate when the transverse axis of one is the conjugate of the other, and the reverse.*

The equation of the conjugate hyperbola, obtained by changing x into y, Art. (108), and a into b, in equation (h), is

$$a^2y^2 - b^2x^2 = a^2b^2.$$

149. The parameter of any diameter of either the ellipse or hyperbola, is a third proportional to the diameter and its conjugate, the conjugate being the mean. Thus, for the parameter of $2a' = \text{B}'\text{A}'$

$$2a' : 2b' :: 2b' : 2p; \quad \text{whence} \quad 2p = \frac{2b'^2}{a'}.$$

The parameter of the transverse axis, $\frac{2b^2}{a}$, is also the parameter of the curve, Art. (115).

For the parameter of the conjugate axis, we have

$$2p = \frac{2a^2}{b}.$$

150. As equations (e') and (h'), Arts. (143), (148), are precisely the same as equations (e) and (h), except that a' and b' enter instead of a and b, it follows that any algebraic expression deduced from the latter, will become the corresponding expression for the former, by changing a into a' and b into b'. Thus the proportions of Arts. (125), (126), become

$$y'^2 : y''^2 :: (a' + x')(a' - x') : (a' + x'')(a' - x''),$$

$$y'^2 : y''^2 :: (x' + a')(x' - a') : (x'' + a')(x'' - a');$$

the first of which shows that, *the squares of the ordinates drawn to any diameter of an ellipse, are to each other as the rectangle of the segments into which the diameter is divided;* and the second that, *the squares of the ordinates drawn to any diameter of an hyperbola, which intersects the curve, are to each other as the rectangles of the distances from the foot of each ordinate, to the vertices of the diameter.*

These properties enable us to construct either curve, having given two conjugate diameters and the angle formed by them. Thus, let A′B′ and D′E′ be two such diameters. Revolve D′E′ until it becomes perpendicular to A′B′; on the two as axes, describe an ellipse (or hyperbola), in which draw any number of ordinates mp, $m'p'$, &c.; then revolve these until they become parallel to the primitive position of D′E′, their extremities will be points of the curve.

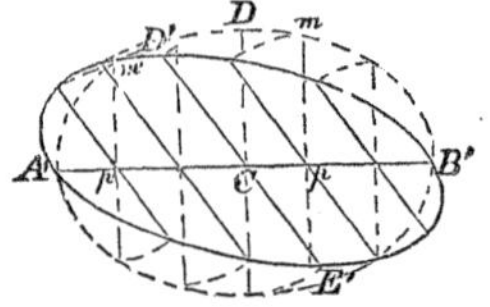

151. The equations of the tangent, Arts. (128), (131), become, when referred to conjugate diameters,

$$a'^2yy'' + b'^2xx'' = a'^2b'^2,$$

$$a'^2yy'' - b'^2xx'' = -a'^2b'^2,$$

the first for the ellipse, and the second for the hyperbola.

152. The equations of condition for supplementary chords Arts. (135), (136), when drawn from the extremities of a diameter $2a'$, become

$$cc' = -\frac{b'^2}{a'^2}, \text{ for the ellipse............(1)},$$

$$cc' = \frac{b'^2}{a'^2}, \text{ for the hyperbola},$$

in which, since the axes of co-ordinates are oblique, c and c' re present the ratio of the sines of the angles which the chords make with the axes, Art. (20).

153. Likewise, equation (1), Art. (137), will belong to a diameter and tangent at its vertex, when referred to conjugate diameters, if we change a into a' and b and b', and regard d and d' as the ratio of the sines, &c.; thus,

$$dd' = -\frac{b'^2}{a'^2}.$$

Comparing this with equation (1) of the preceding article, we have

$$cc' = dd',$$

and the same for the hyperbola. Hence, if $c = d$, $c' = d'$ and the converse. Therefore, in either curve, if a chord, drawn from the extremity of a diameter, is parallel to a tangent, its supplement will be parallel to the diameter passing through the point of contact, and the converse. Also, if one of two supplementary chords is parallel to a diameter, the other will be parallel to its conjugate: or, a set of supplementary chords may always be drawn from the extremities of any diameter parallel to a set of conjugate diameters.

154. The properties of supplementary chords, diameters, and tangents, discussed in the preceding article, give the following constructions.

First. If either the ellipse or hyperbola is traced upon paper; draw any two parallel chords, nn and $n'n'$, and bisect them by a straight line, this will be a diameter, Art. (100). If two diameters be thus constructed, their intersection will be the centre of the curve.

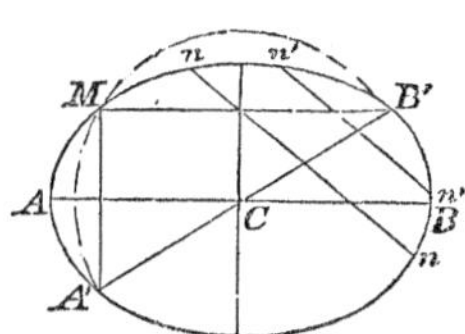

Second. On any diameter, A′B′, found as above, describe a semi-circle and draw two chords from the point M, in which it intersects the curve to the extremities of the diameter; these chords will be supplementary and perpendicular to each other; draw two diameters parallel to these and they will be the axes.

And, in general, to construct a set of conjugate diameters making a given angle with each other. Upon any diameter describe an arc capable of containing the given angle; from the point in which it cuts the curve, draw two chords to the extremities of the diameter; through the centre draw two diameters parallel to these chords; they will be the required diameters.

Third. If one diameter, as D′E′, is given, and it be required to construct its conjugate. From the extremity of any diameter draw a chord Rm, parallel to the given diameter; draw the supplement, Om, of this chord; through the centre draw a diameter CA′ parallel to this supplement, it will be the one required.

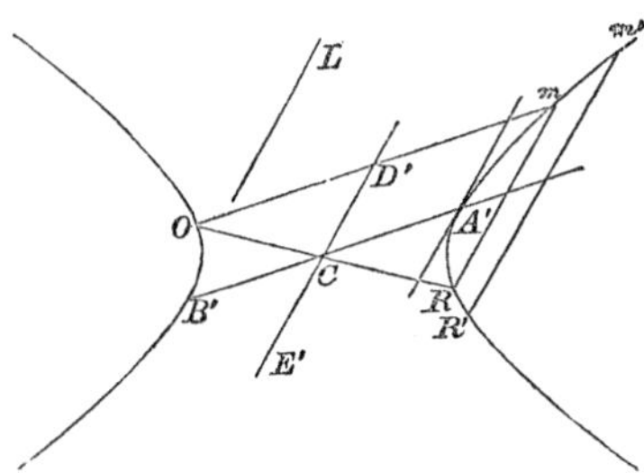

Fourth. To draw a tangent at a given point, as A′. Join this point with the centre; through the extremity O, of any diameter, draw a chord parallel to CA′; draw the supplement of this chord, Rm; parallel to which, draw a line through the given point; it will be the required tangent.

Fifth. To draw a tangent parallel to a given line, as L. Make the same construction as in Art. (137), using any diameter as OR, instead of the transverse axis. Or thus: draw any two chords mR, m′R′, parallel to the given line; bisect them by a straight line; the points A′ and B′, in which this intersects the curve will be the points of contact, through either of which draw a line parallel to the given line, it will be the required tangent.

155. Let CB′ and CD′ be any two semi-conjugate diameters of the ellipse. On the transverse axis AB, describe a semi-circle; through the points B′ and D′ draw two ordinates and produce them to M and M′; draw the radii CM and CM′, and denote the angles MCB and M′CB by β and β'. The right angled triangles CPB′ and CPM, give

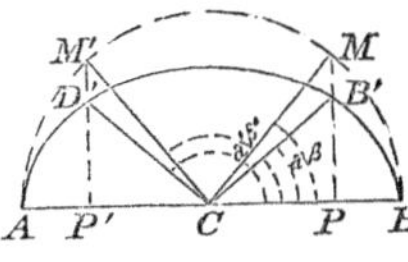

$$\text{tang}\,\alpha \;:\; \text{tang}\,\beta \;::\; \text{PB}' \;:\; \text{PM} \;::\; b \;:\; a,$$

Art. (127); hence

$$\text{tang}\,\alpha = \frac{b}{a}\,\text{tang}\,\beta.$$

Also, the triangles CP′D′ and CP′M′, since the angles at the bases are supplements of α' and β', give

$$\text{tang}\,\alpha' = \frac{b}{a}\,\text{tang}\,\beta'.$$

Multiplying these equations, member by member, we have

$$\text{tang}\,\alpha\,\text{tang}\,\alpha' = \frac{b^2}{a^2}\,\text{tang}\,\beta\,\text{tang}\,\beta' = -\,\frac{b^2}{a^2},$$

Art. (144); hence

$$\text{tang}\,\beta\,\text{tang}\,\beta' = -\,1,$$

and the two radii CM and CM′ are perpendicular to each other, Art. (28); therefore

$$\beta' = 90^\circ + \beta, \qquad \sin\beta' = \cos\beta, \qquad \cos\beta' = -\,\sin\beta.$$

156. From the triangles CPB′ and CPM, we have

$$CP = a' \cos \alpha, \qquad CP = a \cos \beta,$$

$$PB' = a' \sin \alpha, \qquad PM = a \sin \beta;$$

whence, from the first two,

$$a' \cos \alpha = a \cos \beta \ldots\ldots\ldots\ldots(1),$$

and from the second

$$a' \sin \alpha : a \sin \beta :: PB' : PM :: b : a,$$

or

$$a' \sin \alpha = b \sin \beta \ldots\ldots\ldots\ldots(2).$$

In the same way, the triangles CP′D′ and CP′M′ give

$$b' \cos \alpha' = a \cos \beta' = - a \sin \beta \ldots\ldots\ldots(3),$$

$$b' \sin \alpha' = b \sin \beta' = b \cos \beta \ldots\ldots\ldots\ldots\ldots(4),$$

after substituting for $\cos \beta'$ and $\sin \beta'$ their values, as deduced in the preceding article.

Multiplying equations (1) and (4), member by member, and then, (2) and (3), and subtracting the latter product from the former, we have

$$a'b'(\sin \alpha' \cos \alpha - \sin \alpha \cos \alpha') = ab(\cos^2 \beta + \sin^2 \beta),$$

or

$$a'b' \sin(\alpha' - \alpha) = ab \ldots\ldots\ldots\ldots(5).$$

Squaring both members of (1) and (3) and adding, we have

$$a'^2 \cos^2 \alpha + b'^2 \cos^2 \alpha' = a^2.$$

In the same way, from (2) and (4) we obtain

$$a'^2 \sin^2 \alpha + b'^2 \sin^2 \alpha' = b^2.$$

Adding the last two equations, member by member,

$$a'^2 + b'^2 = a^2 + b^2 \ldots\ldots\ldots\ldots(6).$$

157. If we unite equation (1) of Art. (144), with (5) and (6) of the preceding article, we have

$$\text{tang}\,\alpha\,\text{tang}\,\alpha' = -\frac{b^2}{a^2}\ldots\ldots\ldots\ldots(1),$$

$$a'b'\sin(\alpha' - \alpha) = ab\ldots\ldots\ldots\ldots(2),$$

$$a'^2 + b'^2 = a^2 + b^2\ldots\ldots\ldots\ldots(3),$$

three equations containing six quantities, either three of which being given, the others may be determined.

If the angle $\alpha' - \alpha$, made by the conjugate diameters with each other, is given equal to ω, we have

$$\alpha' - \alpha = \omega, \qquad \text{tang}\,\alpha' = \text{tang}\,(\omega + \alpha),$$

and this value in equation (1), will give an equation from which tang α may be found, and thus both α and α', become known.

Let us resume equation (2),

$$a'b'\sin(\alpha' - \alpha) = ab\ldots\ldots\ldots\ldots(2),$$

and at the vertices of any two conjugate diameters A′B′ and D′E′, draw tangents forming a parallelogram. Also at the vertices of the axes, draw tangents forming a rectangle. From the right angled triangle D′RC, we have

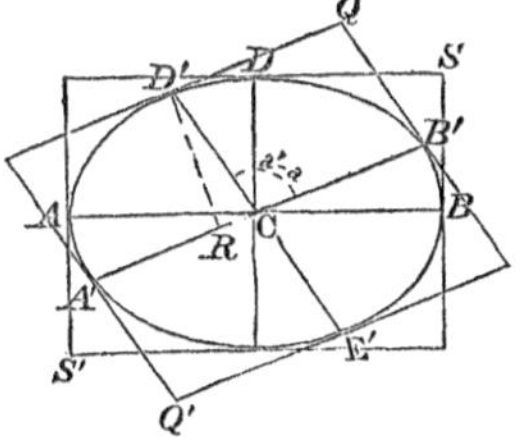

$$\text{D'R} = \text{D'C}\sin\text{D'CR} = b'\sin(\alpha' - \alpha);$$

hence, the first member of equation (2) is equal to CB′ × D′R, or the area of the parallelogram CQ, while the second member is equal to CB × CD, or the area of the rectangle CS. If these equals be multiplied by 4, we have four times the parallelogram CQ, or the parallelogram QQ′, equal to four times the rectangle CS, or the rectangle SS′; that is, *the parallelogram constructed upon any two conjugate diameters of the ellipse, is equivalent to the rectangle upon the axes.*

Multiplying both members of equation (3), by 4, we have

$$4a'^2 + 4b'^2 = 4a^2 + 4b^2.$$

But

$$4a'^2 = (2a')^2, \qquad 4b'^2 = (2b')^2 \text{ \&c.};$$

hence, *the sum of the squares upon any two conjugate diameters of the ellipse, is equal to the sum of the squares upon the axes.*

158. If in equations (1), (2) and (3), of the preceding article, we change b^2 into $-b^2$ and b'^2 into $-b'^2$, we have, for the hyperbola,

$$\text{tang } \alpha \text{ tang } \alpha' = \frac{b^2}{a^2} \ldots\ldots\ldots\ldots (1),$$

$$a'b' \sin(\alpha' - \alpha) = ab \ldots\ldots\ldots\ldots (2),$$

$$a'^2 - b'^2 = a^2 - b^2 \ldots\ldots\ldots\ldots (3),$$

from which either three of the quantities may be determined when the others are known.

If $\alpha' - \alpha = \omega$, is given, α and α' may be determined as in the preceding article.

The perpendicular D′R is equal to

$$\text{D'C} \sin \text{D'CR} = b' \sin(\alpha' - \alpha);$$

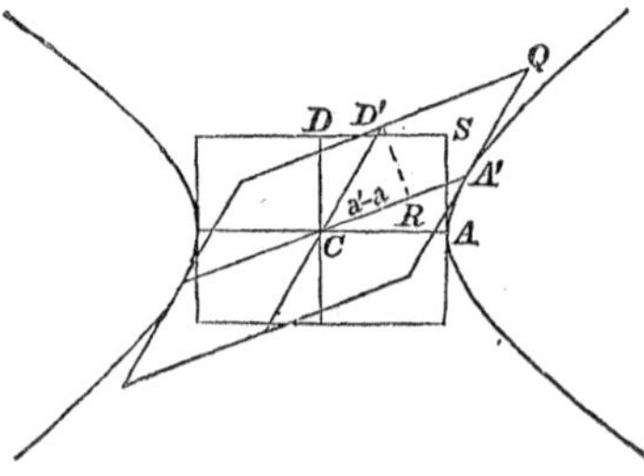

hence, the first member of equation (2) is equal to CA′ × D′R, or the area of the parallelogram CQ; while the second member is equal to the rectangle CS. Multiplying these equals by 4, we obtain *the parallelogram constructed upon any two conjugate diameters of the hyperbola equivalent to the rectangle upon the axes.*

From equation (3), we have

$$4a'^2 - 4b'^2 = 4a^2 - 4b^2,$$

that is, *the difference of the squares of any two conjugate diameters of the hyperbola, is equal to the difference of the squares of the axes.*

159. If the conjugate diameters of an ellipse are equal to each other, the two expressions for a'^2 and b'^2, Art. (143), must be equal, which requires that

$$a^2 \sin^2 \alpha + b^2 \cos^2 \alpha = a^2 \sin^2 \alpha' + b^2 \cos^2 \alpha',$$

and every set of values of α and α' which will satisfy this equation, provided they at the same time satisfy equation (1), Art. (144), will give the position of a set of equal conjugate diameters.

Substituting in the above equation

$$\sin^2 \alpha = 1 - \cos^2 \alpha, \qquad \sin^2 \alpha' = 1 - \cos^2 \alpha',$$

it becomes

$$(a^2 - b^2) \cos^2 \alpha = (a^2 - b^2) \cos^2 \alpha' \ldots\ldots\ldots\ldots (1),$$

which, unless $a = b$, can only be satisfied by making

$$\cos^2 \alpha = \cos^2 \alpha', \qquad \text{or} \qquad \cos \alpha = \pm \cos \alpha'.$$

The first value $\cos \alpha = \cos \alpha'$ gives $\alpha = \alpha'$; hence the two diameters coincide and are not conjugate.

The second value $\cos \alpha = - \cos \alpha'$, satisfies equation (1), Art. (144), and requires the angles to be supplements of each other, or

$$\alpha' + \alpha = 180°;$$

hence, Art. (135), the diameters must be parallel to the supplementary chords drawn from the extremities of the transverse to the extremity of the conjugate axis. They will therefore make a greater angle with each other than any other set of conjugate diameters.

If $a = b$, equation (1), will be satisfied for every set of values of α and α'; hence, in the circle, the conjugate diameters are equal to each other.

When $a' = b'$, equation (e'), Art. (143), becomes

$$a'^2y^2 + a'^2x^2 = a'^4, \quad \text{or} \quad y^2 + x^2 = a'^2,$$

for the ellipse referred to its equal conjugate diameters.

160. By an examination of equation (3), Art. (158), it will be seen that a' can not be equal to b', unless $a = b$; that is, in the hyperbola, there are no equal conjugate diameters, except when the hyperbola is equilateral, in which case each diameter is equal to its conjugate.

OF THE HYPERBOLA REFERRED TO ITS ASYMPTOTES.

161. If in equation (1) of Art. (143), we put $-b^2$ for b^2, it becomes

$$(a^2 \sin^2 \alpha' - b^2 \cos^2 \alpha')y^2 + (a^2 \sin^2 \alpha - b^2 \cos^2 \alpha)x^2$$
$$+ 2(a^2 \sin \alpha \sin \alpha' - b^2 \cos \alpha \cos \alpha')xy = -a^2b^2 \ldots\ldots(1),$$

for the equation of the hyperbola referred to any set of oblique axes having their origin at the centre. We may assign such values to the arbitrary constants α and α', in this equation, as to cause the coefficients of y^2 and x^2 to be 0, and thus have

$$a^2 \sin^2 \alpha' - b^2 \cos^2 \alpha = 0, \qquad a^2 \sin^2 \alpha - b^2 \cos^2 \alpha = 0 \ldots\ldots(2),$$

whence by dividing the first by $a^2 \cos^2 \alpha'$, and the second by

$a^2 \cos^2 \alpha$, and recollecting that $\dfrac{\sin^2}{\cos^2} = \text{tang}^2$, we deduce

$$\text{tang}\ \alpha' = \pm \frac{b}{a}, \qquad \text{tang}\ \alpha = \pm \frac{b}{a}.$$

But it is evident that we can not use $\text{tang}\ \alpha' = \text{tang}\ \alpha$, as such case, the two new axes of co-ordinates would coincide. If, erefore, we take

$$\text{tang}\ \alpha' = \frac{b}{a},$$

we must take

$$\text{tang}\ \alpha = -\frac{b}{a},$$

and the reverse.

These values may be readily constructed, thus: At the vertex A, erect a perpendicular to BA, on which lay off the distances AL and AL′, each equal to b and draw the lines CL and CL′, these will be the new axes of co-ordinates. For the right angled triangles CAL and CAL′ give

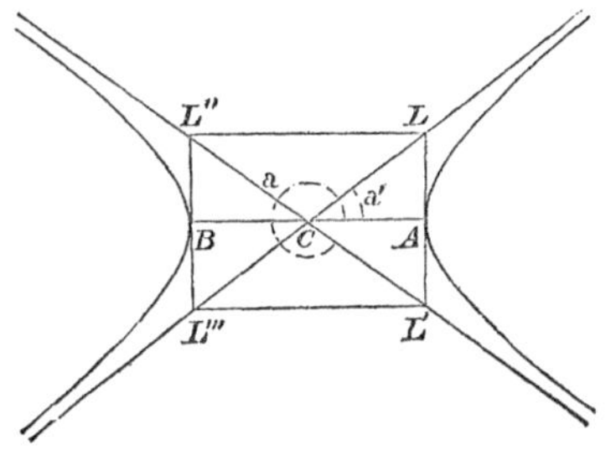

$$\text{tang ACL} = \text{tang ACL}' = \frac{b}{a} = \text{tang}\ \alpha'.$$

But the tangent of ACL′, taken with a contrary sign, is equal to the tangent of $360° - \text{ACL}'$, and also equal to $-\frac{b}{a} = \text{tang}\ \alpha$; hence

$$360° - \text{ACL}' = \alpha,$$

and CL′ is the new axis of X, and CL the new axis of Y, the angles α and α' being as marked on the figure.

Since

$$\text{CL} = \sqrt{a^2 + b^2},$$

the right angled triangle ACL gives

$$\sin\alpha' = \frac{b}{\sqrt{a^2 + b^2}}, \qquad \cos\alpha' = \frac{a}{\sqrt{a^2 + b^2}},$$

and since

$$\sin\alpha' = -\sin\alpha, \qquad \cos\alpha' = \cos\alpha,$$

we also have

$$\sin\alpha = \frac{-b}{\sqrt{a^2 + b^2}}, \qquad \cos\alpha = \frac{a}{\sqrt{a^2 + b^2}}.$$

Substituting these values, together with conditions (2), in equation (1), we have

$$2\left(\frac{-a^2b^2}{a^2 + b^2} - \frac{a^2b^2}{a^2 + b^2}\right)xy = -a^2b^2,$$

whence

$$xy = \frac{a^2 + b^2}{4}, \qquad \text{or} \qquad xy = m \ldots\ldots\ldots\ldots (3),$$

placing m for $\dfrac{a^2 + b^2}{4}$.

Solving this equation, we have

$$y = \frac{m}{x},$$

in which, as x increases, y diminishes ; when x becomes infinite y becomes 0 ; and as y can be negative for no positive value of x, it follows that the axis of X, or the line CL′, continually approaches the curve and touches it at an infinite distance without cutting it. By solving the equation with reference to x, it may be proved that the line CL enjoys the same property. These two lines are called asymptotes of the hyperbola ; and in general, *an asymptote of a curve is a line, which continually approaches the curve and becomes tangent to it at an infinite distance.*

By an inspection of the figure, it is readily seen that the asymptotes of the hyperbola are the diagonals of the rectangle described on the axes.

Equation (3) is called *the equation of the hyperbola referred to its centre and asymptotes.*

If the hyperbola is equilateral, $\alpha' = 45°$, $\tan \alpha' = 1$, the angle $LCL' = 90°$, and the asymptotes are perpendicular to each other.

162. If we take the expression, Art. (132),

$$x = \frac{a^2}{x''} = CT,$$

and make $x'' = a$, the least value it can have for points of the curve, Art. (107), we shall obtain

$$CT = a = CA,$$

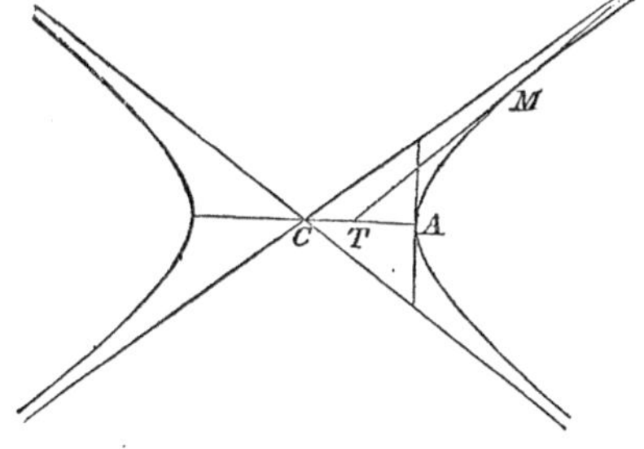

which is the greatest value of CT. As x'' increases, CT diminishes until $x'' = \infty$, when $CT = 0$, its least value, and the tangent coincides with the asymptote ; hence all tangents drawn to the hyperbola intersect the transverse axis between the centre and the vertex of that branch to which they are drawn; and *the asymptotes are the limits of all tangents.*

163. If we multiply both members of equation (3), Art. (161), by sin $2\alpha'$, the sine of the angle LCL′ included by the asymptotes, we have

$$xy \sin 2\alpha' = m \sin 2\alpha'.$$

The second member of this equation is constant, and the first for any point of the curve, as M, is

$$\text{CP} \times \text{PM} \sin \text{MP}n = \text{CP} \times \text{M}n,$$

which is the area of the parallelogram CPMR. Hence, the areas of all parallelograms described on the abscissas and ordinates of points of the curve, referred to the asymptotes, are equal, each being measured by the expression $m \sin 2\alpha'$.

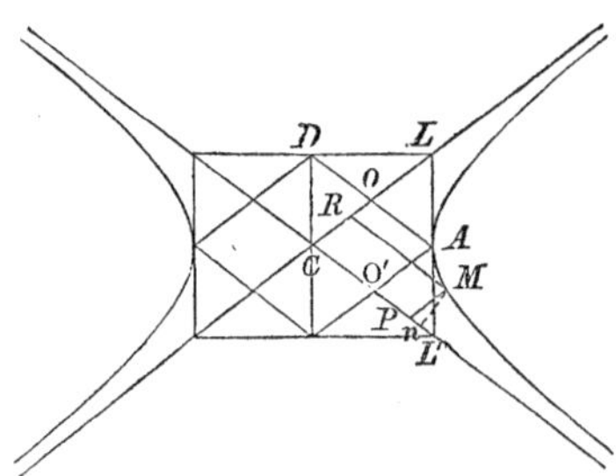

If the point M is placed at the vertex A, the parallelogram becomes the rhombus AOCO′, each of its sides being

$$\frac{1}{2}\,\text{CL} = \frac{1}{2}\sqrt{a^2 + b^2}.$$

This rhombus, described on the abscissa and ordinate of the vertex, is called *the power of the hyperbola,* is equivalent to the parallelogram described on the abscissa and ordinate of any point of the curve, and as is readily seen from the figure, is *one eighth* of the rectangle described on the axes.

In the equilateral hyperbola

$$2\alpha' = 90°, \qquad \sin 2\alpha' = 1,$$

and the power becomes a square, the area of which is m.

164. Let x'' y'' denote the co-ordinates of any point, as M, of the hyperbola. The equation of a right line passing through this point, will be of the form

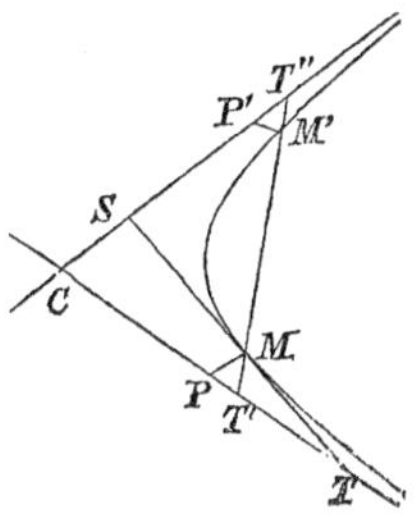

$$y - y'' = d(x - x'') \ldots\ldots\ldots\ldots (1).$$

The equation of the hyperbola being

$$xy = m \ldots\ldots\ldots\ldots (2),$$

the condition that the point M shall be on the curve, will be

$$x''y'' = m.$$

Subtracting this from equation (2), we have

$$xy - x''y'' = 0.$$

Combining this with equation (1), by substituting the value of y taken from (1), we obtain

$$y''(x - x'') + dx(x - x'') = 0, \quad \text{or} \quad (x - x'')(y'' + dx) = 0.$$

Placing the two factors of the last equation, separately, equal to 0, we obtain for all the common points, Art. (128,)

$$x - x'' = 0 \qquad \text{or} \qquad x = x'',$$

$$y'' + dx = 0 \qquad \text{or} \qquad x = -\frac{y''}{d} \ldots\ldots\ldots\ldots (3).$$

The first value of x is evidently the abscissa of the point M, the second must then be that of M′.

Now if the line MM′ be revolved about M until the point M′ coincides with M, it will become a tangent, and the value of x in equation (3) will become x'', whence we have

$$d = -\frac{y''}{x''}.$$

This value, in equation (1), gives

$$y - y'' = -\frac{y''}{x''}(x - x''),$$

for *the equation of the tangent, referred to the asymptotes.*

If in this equation we make $y = 0$, we have

$$x - x'' = x'' = \text{PT},$$

that is, *the subtangent is equal to the abscissa of the point of contact.*

And, since PT = CP, we have MT = MS; that is, *the part of the tangent included between the asymptotes is bisected at the point of contact.*

If in equation (1) we make $y = 0$, we find

$$x - x'' = -\frac{y''}{d} = \text{CT}' - \text{CP} = \text{PT}',$$

the same value found in equation (3) for the abscissa of M′; hence

$$\text{PT}' = \text{M}'\text{P}',$$

and the two triangles MPT′ and M′P′T″, having their angles also equal, are equal, and MT′ = M′T″; that is, *if any straight line be drawn cutting the hyperbola, the parts included between the curve and asymptotes will be equal.*

This property enables us to construct the hyperbola by points when a single point and the asymptotes are given. Through the point, as M, draw any right line; from the point in which it cuts one of the asymptotes, lay off the distance T″M′, equal to the distance MT′, from the given point to that in which it cuts the other asymptote; the extremity of this distance will be a point of the curve.

165. If a tangent TS be drawn at any point, M, of the hyperbola, and the half tangent MT be denoted by t, the half diameter MC by a', and the perpendicular Mn be drawn, the two right angled triangles MnC and MnT will give

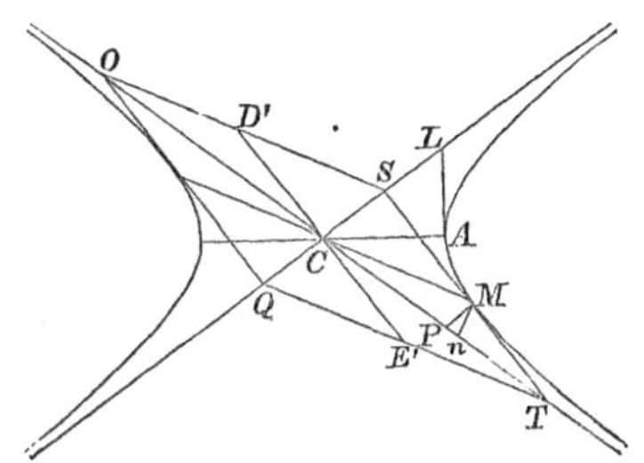

$$a'^2 = (x + Pn)^2 + \overline{Mn}^2,$$

$$t^2 = (x - Pn)^2 + \overline{Mn}^2.$$

Subtracting and reducing, we have

$$a'^2 - t^2 = 4xPn.$$

But the right angled triangle MPn, in which the angle MPn $= 2\alpha'$, gives

$$Pn = y \cos 2\alpha',$$

hence

$$a'^2 - t^2 = 4xy \cos 2\alpha' = 4m \cos 2\alpha'.$$

The second member of this equation is constant, and will therefore be the same for any position of the point M. If this point be placed at the vertex A, the half diameter will be CA $= a$, and the half tangent AL $= b$; hence

$$a^2 - b^2 = 4m \cos 2\alpha' ;$$

whence

$$a'^2 - t^2 = a^2 - b^2.$$

But, Art. (158), we have

$$a'^2 - b'^2 = a^2 - b^2,$$

therefore, we must have

$$t = b' \qquad \text{or} \qquad 2t = 2b' ;$$

that is, *if a tangent be drawn at any point of the hyperbola, the part intercepted between the asymptotes will be equal to the conjugate of the diameter passing through the point of contact.*

Since the line E′D′ is equal and parallel to TS = QO, it follows that the figure QOST is a parallelogram, and that the vertices of any parallelogram described on a set of conjugate diameters, will lie on the asymptotes.

OF THE POLAR EQUATIONS OF THE ELLIPSE AND HYPERBOLA.

166. If in equation (e), Art. (105), we substitute the values of x and y from formulas (2) of Art. (69),

$$x = a' + r \cos v, \qquad y = b' + r \sin v,$$

we shall obtain after reduction

$$(a^2 \sin^2 v + b^2 \cos^2 v)r^2 + 2(a^2b' \sin v + b^2a' \cos v)r$$
$$+ a^2b'^2 + b^2a'^2 - a^2b^2 = 0 \ldots\ldots (1),$$

for *the general polar equation of the ellipse.*

By changing b^2 into $- b^2$, this will become the general polar equation of the hyperbola.

By assigning particular values to a' and b', in the above equation, the pole may be placed at any point in the plane of the curve.

First. If the pole is on the curve, we must have, Art. (109),

$$a^2b'^2 + b^2a'^2 - a^2b^2 = 0,$$

and the equation reduces to

$$[(a^2 \sin^2 v + b^2 \cos^2 v)r + 2(a^2b' \sin v + b^2a' \cos v)]r = 0,$$

which may be satisfied by placing $r = 0$, or

$$(a^2 \sin^2 v + b^2 \cos^2 v)r + 2(a^2b' \sin v + b^2a' \cos v) = 0 \ldots\ldots(2).$$

The pole being on the curve, one value of r is necessarily equal to 0, and the other deduced from equation (2), will, for each value of v, give the distance from the pole to the second point, in which the radius vector cuts the curve, Art. (70).

If this second value of r becomes 0, the radius vector will become tangent to the curve, and equation (2) will reduce to

$$a^2b' \sin v + b^2a' \cos v = 0,$$

or

$$\frac{\sin v}{\cos v} = \text{tang } v = -\frac{b^2a'}{a^2b'},$$

as it should be, Art. (128).

For the hyperbola, we shall have the same discussion and result, except that $-b^2$ takes the place of b^2.

Second. If the pole be placed at the centre, we have

$$a' = 0, \qquad b' = 0,$$

which reduces equation (1) to

$$(a^2 \sin^2 v + b^2 \cos^2 v)r^2 - a^2b^2 = 0;$$

whence

$$r = \pm \sqrt{\frac{a^2b^2}{a^2 \sin^2 v + b^2 \cos^2 v}} \ldots\ldots\ldots (3).$$

The second value of r is negative for all values of v, and therefore gives no point of the curve, Art. (69).

The first value is positive for all values of v, and as v varies from 0 to 360°, will give all points of the curve.

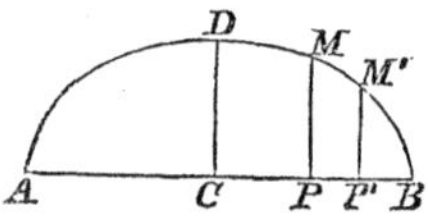

If $v = 0$, $\sin v = 0$, $\cos v = 1$, and r reduces to

$$r = a = \text{CB}.$$

If $v = 90°$, $\sin v = 1$, $\cos v = 0$, and

$$r = b = \text{CD}.$$

If in the first value of r, (3), we put for $\sin^2 v$ its value $1 - \cos^2 v$, divide both terms of the fraction under the radical sign by a^2, and then place

$$\frac{a^2 - b^2}{a^2} = e^2,$$

e representing the eccentricity of the ellipse, Art. (118), we shall obtain

$$r = \frac{b}{\sqrt{1 - e^2 \cos^2 v}}.$$

For the hyperbola, equation (3), becomes

$$r = \pm \sqrt{\frac{-a^2b^2}{a^2 \sin^2 v - b^2 \cos^2 v}};$$

the second value of which is negative for all values of v.

The first value is positive but imaginary, unless the denominator is negative which requires

$$a^2 \sin^2 v - b^2 \cos^2 v < 0, \qquad \text{or} \qquad \text{tang}\, v < \pm \frac{b}{a}.$$

If $v = 0$, we have, as above

$$r = a = \text{CA}.$$

As v increases from 0, the denominator will be negative until

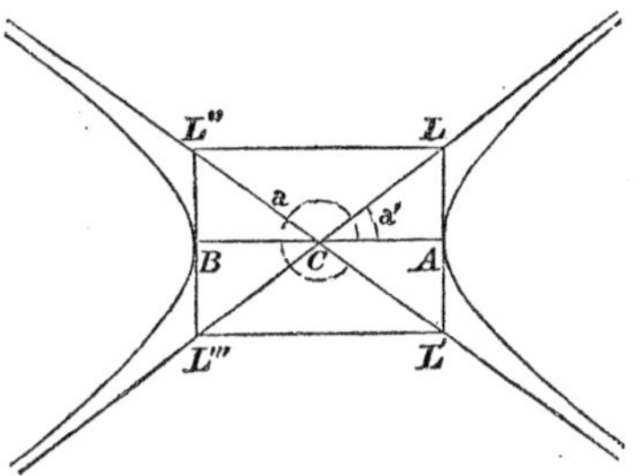

$$a^2 \sin^2 v = b^2 \cos^2 v, \qquad \text{tang } v = \frac{b}{a},$$

when the value of r will be infinite, in which case $v = \text{LCA}$, and the radius vector coincides with the asymptote CL, Art. (161). As v increases beyond this value, $a^2 \sin^2 v$ becomes greater than $b^2 \cos^2 v$, and r will be imaginary until

$$\text{tang } v = -\frac{b}{a},$$

in which case $v = \text{ACL}''$ and the radius vector coincides with CL''.

When $v = 180°$, we have

$$r = a = \text{CB},$$

and as v still increases, we shall continue to have real values for r until it coincides with CL''', when tang v again becomes equal to $\frac{b}{a}$, and from this point the values of r will be imaginary until the radius vector coincides with CL', when they again become real and continue so to $v = 360°$.

The first value of r thus gives all the points in both branches of the hyperbola.

By a process similar to that pursued in the ellipse, the first value of r may be reduced to

$$r = \frac{b}{\sqrt{e^2 \cos^2 v - 1}},$$

in which e represents the eccentricity of the hyperbola, Art. (119).

Third. If the pole be placed at the right hand focus, for which

$$a' = \sqrt{a^2 - b^2} = c, \qquad b' = 0,$$

equation (1), becomes

$$(a^2 \sin^2 v + b^2 \cos^2 v)r^2 + 2b^2c \cos vr - b^4 = 0.$$

If for $\sin^2 v$ we put its value $1 - \cos^2 v$, and for $a^2 - b^2$, its value c^2, this equation reduces to

$$(a^2 - c^2 \cos^2 v)r^2 + 2b^2c \cos vr = b^4,$$

from which

$$r = \frac{-b^2c \cos v}{a^2 - c^2 \cos^2 v} \pm \sqrt{\frac{b^4}{a^2 - c^2 \cos^2 v} + \frac{b^4c^2 \cos^2 v}{(a^2 - c^2 \cos^2 v)^2}},$$

or reducing

$$r = \frac{-b^2c \cos v \pm ab^2}{a^2 - c^2 \cos^2 v} = \frac{\pm ab^2 - b^2c \cos v}{(a + c \cos v)(a - c \cos v)};$$

whence, the two values

$$r = \frac{b^2}{a + c \cos v} \ldots\ldots(4), \qquad r = \frac{-b^2}{a - c \cos v} \ldots\ldots(5).$$

Since for the ellipse

$$c = \sqrt{a^2 - b^2} < a \quad \text{and} \quad \cos v < 1,$$

the second value of r is always negative and must therefore be rejected.

As v varies from 0 to 360°, the first value of r will be positive, and give all points of the ellipse.

For the hyperbola, expressions (4) and (5) become

$$r = \frac{-b^2}{a + c \cos v} \ldots\ldots(6), \qquad r = \frac{b^2}{a - c \cos v} \ldots\ldots(7).$$

The first value of r will be positive whenever the denominator is negative. But this can not be unless $\cos v$ is negative and numerically greater than $\frac{a}{c}$. Every value of v, beginning with 0, will then make r negative until

$$\cos v = -\frac{a}{c} = -\frac{a}{\sqrt{a^2 + b^2}},$$

when the radius vector will be parallel to the asymptote CL″, Art. (161), and r will be infinite. As v now increases, $\cos v$ will increase numerically until $v = 180°$, when $\cos v = -1$, and

$$r = \frac{b^2}{c - a},$$

which is positive, and gives the vertex B. As v increases from this point, $\cos v$ will diminish numerically and r will be positive until we again have

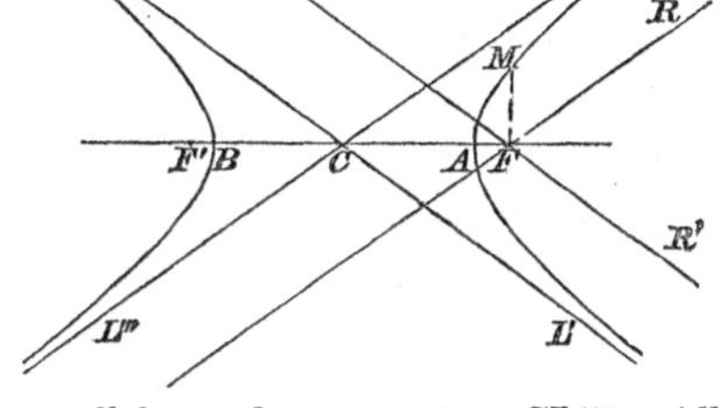

$$\cos v = -\frac{a}{c},$$

when $r = \infty$, and the radius vector becomes parallel to the asymptote CL‴. All values of v not included within these limits will make the first value of r negative and give no points of the curve. Thus, it appears that this value of r gives all the points in the left hand branch of the hyperbola, and no others.

The second value of r will be positive, when the denominator is positive. Commencing with $v = 0$ $\cos v = 1$, we have

$$r = \frac{b^2}{a - c},$$

which is negative. As v increases, cos v diminishes, and r will remain negative until $a = c \cos v$, when

$$\cos v = \frac{a}{c} = \frac{a}{\sqrt{a^2 + b^2}},$$

r reduces to infinity, and the radius vector takes the position FR, parallel to the asymptote CL. As v increases from this point, r will be positive until it takes the position FR′ parallel to CL′.

When $v = 90°$, $\cos v = 0$ and

$$r = \frac{b^2}{a} = \text{FM}.$$

When $v = 180°$, $\cos v = -1$, and

$$r = \frac{b^2}{a + c} = \text{FA}.$$

The second value of r, therefore, gives all the points in the right hand branch, and no others.

If in expressions (4), (6) and (7), we put $-c$ for c, the pole, in each case, will be placed at the left hand focus.

If the eccentricity of an ellipse be denoted by e, we have, Art. (118),

$$e = \frac{c}{a}, \qquad e^2 = \frac{c^2}{a^2} = \frac{a^2 - b^2}{a^2},$$

from which, we deduce

$$c = ae, \qquad b^2 = a^2(1 - e^2) \quad \ldots\ldots\ldots\ldots (8).$$

Substituting these values in expression (4), we find

$$r = \frac{a(1 - e^2)}{1 + e \cos v},$$

which expresses the value of r in terms of the eccentricity of the ellipse.

For the hyperbola, Art. (119), we have

$$c = ae, \qquad -b^2 = a^2(1 - e^2) \ldots\ldots\ldots\ldots (9).$$

These values in expressions (6) and (7), give

$$r = \frac{a(1 - e^2)}{1 + e \cos v}, \qquad r = -\frac{a(1 - e^2)}{1 - e \cos v},$$

in terms of the eccentricity of the hyperbola.

From equations (8) and (9) we deduce the numerical value

$$a(1 - e^2) = \frac{b^2}{a};$$

hence, the numerator of each of the above values of r is equal to *one half the parameter of the curve*, Art. (149); as is also the case in the parabola, Art. (104).

DISCUSSION OF THE GENERAL EQUATION OF THE SECOND DEGREE.

167. Every equation of the second degree between two variables, must be a particular case of the most general form

$$ay^2 + bxy + cx^2 + dy + ex + f = 0 \ldots\ldots\ldots\ldots (1),$$

which, by assigning particular values to the constants a, b, c, &c., may be made to represent every line of the second order, Art. (33).

Although there are six terms in the above equation, and apparently six arbitrary constants, yet it must be observed that both members of the equation may be divided by the coefficient of either term, as a, thus reducing it to the form

$$y^2 + b'xy + c'x^2 + d'y + e'x + f' = 0,$$

in which there are but *five* constants, and to which we can assign but *five* arbitrary conditions.

In order then to estimate the number of *arbitrary constants* in any general equation, or equation given in form only, we divide by the coefficient of one of the terms, and then count the number of different coefficients remaining. This will indicate the number of arbitrary conditions which the given equation may be made to fulfil.

In commencing the discussion of equation (1), we may regard the axes of co-ordinates, to which the line represented by it is referred, as at right angles; for if they were oblique, a change of reference might be made by means of formulas (5), Art. (67), and a new equation of precisely the same form would evidently result.

168. By solving equation (1), of the preceding article, with reference to y, we obtain

$$y = -\frac{bx+d}{2a} \pm \frac{1}{2a}\sqrt{(b^2-4ac)x^2+2(bd-2ae)x+d^2-4af} \ldots (1),$$

from which we may readily construct the line by points as in Art. (22). Each value of x which will make the quantity under the radical sign positive, will give two real values for y, and correspond to two points of the curve. These points may be constructed by laying off from A as an origin, the assumed value of x, as AP; at P erect the perpendicular PM, on which lay off

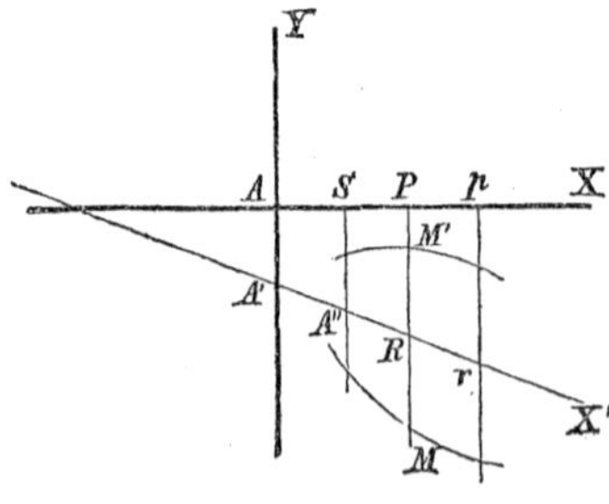

$$PR = -\frac{bx+d}{2a};$$

from R lay off RM′ in the positive direction of the ordinates, and RM in the negative, each equal to the second part of the value of

y; PM$'$ will be represented by the first, and PM by the second value of y, and M$'$ and M will be the corresponding points of the curve.

Since the point R is midway between the two points M and M$'$, it follows that the chord MM$'$ is bisected at R. But since the points R, r, &c., are constructed by laying off the different values of the expression

$$-\frac{bx+d}{2a},$$

it follows that they must all lie on the right line whose equation is

$$y = -\frac{bx+d}{2a}, \qquad \text{or} \qquad y = -\frac{bx}{2a} - \frac{d}{2a};$$

hence, this line will bisect all chords drawn parallel to the axis of Y; *it is therefore a diameter of the curve*, Art. (100), and may at once be constructed by laying off AA$' = -\frac{d}{2a}$, and through A$'$, drawing the line A$'$R, making with AX an angle whose tangent is $-\frac{b}{2a}$, Art. (26).

Hence, *if an equation of the second degree be solved with reference to y, the first member placed equal to that part of the second, which does not contain the radical, will give the equation of a diameter bisecting chords parallel to the axis of* Y.

If the equation be solved with reference to x, a similar discussion will show that, *the first member, placed equal to that part of the second which does not contain the radical, will give the equation of a diameter bisecting chords parallel to the axis of* X.

169. If in equation (1) of the preceding article, we place

$$bd - 2ae = m, \qquad d^2 - 4af = n,$$

it becomes

$$y = -\frac{bx+d}{2a} \pm \frac{1}{2a}\sqrt{(b^2-4ac)x^2+2mx+n} \ldots\ldots(1).$$

Let us now remove the origin of co-ordinates to the point A′, A′R being taken as a new axis of abscissas, the new axis of ordinates being the same as the primitive, A′Y. In the formulas (3), Art. (67), we must then have

$$a' = 0, \qquad b' = -\frac{d}{2a}, \qquad \alpha' = 90^\circ, \qquad \text{tang}\ \alpha = -\frac{b}{2a},$$

and the formulas become

$$x = x' \cos \alpha, \qquad y = -\frac{d}{2a} + x' \sin \alpha + y',$$

or by substituting for x' in the second, its value taken from the first, we have

$$x = x' \cos \alpha, \qquad y = -\frac{bx+d}{2a} + y'.$$

Substituting these values of x and y in equation (1), squaring both members and clearing the denominators, we have

$$4a^2y'^2 = (b^2-4ac)\cos^2 \alpha x'^2 + 2m \cos \alpha x' + n \ldots\ldots\ldots\ldots(2),$$

and this may be put under the form

$$4a^2y'^2 = (b^2-4ac)\cos^2 \alpha \left(x'^2 + \frac{2mx'}{(b^2-4ac)\cos \alpha}\right) + n.$$

If to the quantity within the parenthesis, we add

$$\frac{m^2}{(b^2-4ac)^2 \cos^2 \alpha},$$

it will be a perfect square. To preserve the equality, we must then subtract, without the parenthesis, the same expression multiplied by $(b^2 - 4ac)\cos^2 \alpha$, or

$$\frac{m^2}{b^2 - 4ac}.$$

The above equation then takes the form

$$4a^2y'^2 = (b^2 - 4ac)\cos^2\alpha\left(x' + \frac{m}{(b^2-4ac)\cos\alpha}\right)^2 + n - \frac{m^2}{b^2-4ac} \dots (3),$$

If now we again transfer the origin to a point, as A″, on the axis of X′, at a distance from A′ equal to

$$-\frac{m}{(b^2 - 4ac)\cos\alpha} \dots\dots\dots (3'),$$

the axis of abscissas remaining the same and the new axis of ordinates being parallel to A′Y, the formulas (2), Art. (67), become

$$x' = -\frac{m}{(b^2 - 4ac)\cos\alpha} + x'', \qquad y' = y''.$$

Substituting these in equation (3), and transposing, we have

$$4a^2y''^2 - (b^2 - 4ac)\cos^2\alpha x''^2 = n - \frac{m^2}{b^2 - 4ac} \dots\dots\dots (4),$$

for the equation of the line referred to the two axes A″X′ and A″S.

If $b^2 - 4ac$ *is negative;* the essential sign of the second term of the first member will be positive. If the second member is also positive, the equation will be of the same form as equation (e'), Art. (143), and will therefore represent an ellipse referred to its centre and conjugate diameters.

If the second member is negative, the equation will indicate that the sum of two positive quantities is negative, and can be satisfied by no values of x'' and y''. The line represented by it is then said to be imaginary, and an *imaginary ellipse is a particular case of the ellipse.*

If the second member of the equation is 0, it will indicate that the sum of two positive quantities is equal to 0, and can be satisfied by no values, except

$$y'' = 0, \qquad x'' = 0,$$

which, Art. (16), are the equations of *a point, also a particular case of the ellipse.*

If $b = 0$ and $a = c$, we have

$$\text{tang } \alpha = 0, \qquad \cos \alpha = 1,$$

and equation (4) will reduce to the form

$$y''^2 + x''^2 = R^2,$$

which is the equation of a circle, Art. (35), *another particular case of the ellipse.*

If $b^2 - 4ac$ *is positive;* and the second member negative, equation (4) will be of the same form as equation (h'), Art. (148). If the second member is positive, the signs of all the terms may be changed and it will still be of the same form, x'' having the place of y, and y'' the place of x, Art. (108). In either case it will therefore represent an hyperbola referred to its centre and conjugate diameters.

If the second member is 0, the equation may be solved with reference to y''^2, and will take the form

$$y''^2 = r'^2 x''^2, \quad \text{or} \quad y'' = \pm\, r'x'',$$

representing two right lines which intersect; *a particular case of the hyperbola.*

If $b = 0$ and $a = -c$, the equation takes the form

$$y''^2 - x''^2 = -R^2,$$

the equation of an equilateral hyperbola, Art. (112); *another particular case of the hyperbola.*

If $b^2 - 4ac = 0$, the expression (3′) will be infinite, and the transformation made on equation (3) becomes impossible, but under this supposition equation (2) reduces to

$$4a^2y'^2 = 2m \cos \alpha x' + n \ldots\ldots\ldots\ldots(5).$$

If in this, we transfer the origin to a point on the axis of X′, at a distance from A′, equal to

$$- \frac{n}{2m \cos \alpha} \ldots\ldots\ldots\ldots(6),$$

the formulas of Art. (67) will become

$$x' = - \frac{n}{2m \cos \alpha} + x'', \qquad y' = y'',$$

and these substituted in equation (5), give

$$4a^2y''^2 = 2m \cos \alpha x'', \qquad \text{or} \qquad y''^2 = \frac{2m \cos \alpha}{4a^2} x'',$$

which is of the same form as equation (7), Art. (99), and therefore represents a parabola.

If $m = 0$, expression (6) will be infinite and the last transformation become impossible, but in this case equation (2) becomes

$$4a^2y'^2 = n, \qquad \text{or} \qquad y' = \pm \frac{1}{2a}\sqrt{n},$$

x' being indeterminate, and will represent *two right lines parallel to the axis of* X′, when n is positive; *one right line which coincides with the axis of* X′ when $n = 0$, Art. (21); and *two imaginary right lines* when n is negative. *These are particular cases of the parabola.*

170. The above discussion evidently depends upon the fact that the given equation contains the second power of y, or that a is not 0.

If $a = 0$ and c is not, the equation may be solved with reference to x, and the same results deduced in precisely the same manner.

If $a = 0$ and $c = 0$, and b is not, the general equation takes the form

$$bxy + dy + ex + f = 0 \ldots\ldots\ldots\ldots(1).$$

Let us now, by the aid of the general formulas, Art. (67),

$$x = a' + x', \qquad y = b' + y',$$

change the origin of co-ordinates, without changing the direction of the axes. We thus obtain

$$bx'y' + (a'b + d)y' + (b'b + e)x' + a'b'b + b'd + a'e + f = 0 \ldots(2).$$

In this equation we have two arbitrary constants, a' and b', and may therefore assign such values to them as to give

$$a'b + d = 0 \qquad b'b + e = 0,$$

or

$$a' = -\frac{d}{b}, \qquad b' = -\frac{e}{b}.$$

Substituting these values in equation (2), it reduces to

$$bx'y' - \frac{de}{b} + f = 0, \quad \text{or} \quad x'y' = \frac{de - bf}{b^2},$$

which, since the axes of co-ordinates are at right angles to each other, is the equation of an equilateral hyperbola referred to its centre and asymptotes, Art. (161). Equation (1) then represents the same hyperbola, referred to two right lines parallel to its asymptotes.

If $a = 0$, $b = 0$, $c = 0$, the equation ceases to be an equation of the second degree.

From the previous discussion, we conclude, *that every equation of the second degree between two variables represents one of the conic sections, that is, either a parabola, an ellipse or hyperbola, or one of their particular cases.*

A parabola when $b^2 - 4ac = 0.$

An ellipse when $b^2 - 4ac < 0.$

An hyperbola when $b^2 - 4ac > 0.$

The parabola when $b^2 - 4ac = 0.$

171. Under this supposition, the value of y, equation (1), Art. (169), reduces to

$$y = -\frac{bx + d}{2a} \pm \frac{1}{2a}\sqrt{2mx + n} \dots\dots\dots\dots (1).$$

Every value of x, which will make the quantity under the radical sign positive, will give two real values of y and two corresponding points of the curve.

The value of x, which makes this quantity 0, will give two equal values of y, the two corresponding points unite, and the ordinate produced is tangent to the curve, Art. (35).

Every value of x, which makes the quantity under the radical sign negative, gives imaginary values for y and no points of the curve.

If we place

$$2mx + n = 0, \qquad \text{we have} \qquad x = -\frac{n}{2m},$$

which is the only value of x that will reduce the quantity under the radical sign to 0. Denoting this value by x', the value of y may be written

$$y = -\frac{bx + d}{2a} \pm \frac{1}{2a}\sqrt{2m(x - x')} \dots\dots\dots\dots (2),$$

since

$$2mx + n = 2m\left(x + \frac{n}{2m}\right).$$

Now, *if m is positive;* whether x' be positive or negative, every value of $x > x'$ will give two real and unequal values for y; $x = x'$ will give two equal values; and every value of $x < x'$ will give imaginary values. Hence, the curve extends indefinitely in the direction of x positive, is tangent to the ordinate PV, which corresponds to the abscissa x', and has no points on the left of this ordinate, as indicated by the full line in the figure.

If m is negative, every value of $x > x'$ will give imaginary values for y; $x = x'$ will give equal values, and $x < x'$ will give two real and unequal values. Hence, the curve is limited in the direction of x positive by the produced ordinate PV, and extends indefinitely in the direction of x negative, as indicated by the dotted line in the figure. Hence, in order to obtain the limit of the curve in the direction of the axis of abscissas, *we solve its equation with reference to y, place the quantity under the radical sign equal to* 0, *and deduce the value of x, this value will be the abscissa of the limit, lay it off and through its extremity draw a line parallel to the axis of* Y, *it will be the limit;* and this limit will be tangent to the curve at the vertex of that diameter which bisects the chords parallel to the axis of Y.

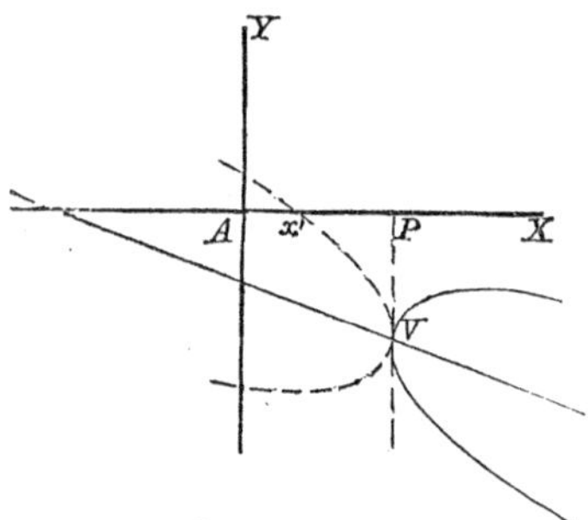

If the coefficient of x under the radical sign is positive, the curve will lie entirely on the right of this limit; if negative, on the left.

By solving the equation with reference to x, we may, in a similar way, construct the limit in the direction of the axis of Y.

If $m = 0$, the value of y, equation (1), becomes

$$y = -\frac{bx + d}{2a} \pm \frac{1}{2a}\sqrt{n},$$

or

$$y = -\frac{bx}{2a} - \frac{d}{2a} + \frac{\sqrt{n}}{2a}, \qquad y = -\frac{bx}{2a} - \frac{d}{2a} - \frac{\sqrt{n}}{2a},$$

the equations of *two parallel straight lines*, Art. (28), when n is positive; which reduce to *one straight line*, when $n = 0$; and to *two imaginary parallels*, when n is negative, as seen in Art. (169). Hence, an equation of a parabola being solved with reference to either variable, if the quantity under the radical sign is *a positive constant*, the equation will represent *two parallel straight lines.*

If this quantity *is* 0, or the radical disappears, the equation will represent *one straight line.* If this quantity *is a negative constant*, the equation will represent *two imaginary parallels.*

It may be remarked, that in the first case, the line whose equation is

$$y = -\frac{bx + d}{2a},$$

bisects all chords, terminated in the two lines and parallel to the axis of Y, and therefore strictly fulfils the condition of a diameter, Art. (100).

In the second case, the line represented by the equation is the diameter itself.

In the third case, the diameter may be constructed while the lines do not exist.

172. By solving the equation with reference to x, we find for the equation of the diameter which bisects all chords parallel to the axis of X; Art. (168),

$$x = -\frac{by + e}{2c}; \qquad \text{whence} \qquad y = -\frac{2cx}{b} - \frac{e}{b};$$

but since $b^2 - 4ac = 0$, we have

$$\frac{b}{2a} = \frac{2c}{b},$$

hence the coefficient of x in the above equation is equal to the coefficient of x in the equation

$$y = -\frac{bx}{2a} - \frac{d}{2a},$$

and the two diameters represented by these equations are parallel, Art. (28).

173. By an application of the foregoing principles we are enabled to represent on paper, a parabola whose equation is given, without taking the trouble to determine many of its points.

First, find the points in which the curve cuts the axes of coordinates, Art. (22); then solve the equation with reference to each variable in succession, and construct the diameters which bisect the chords parallel to the axes, Arts. (168), (26); then construct the limits of the curve in the direction of both axes, Art. (171); and draw a curve tangent to these limits at the points at which they intersect the diameters and through the points first determined, taking care to make it symmetrical with respect to both of the diameters.

Examples.

First, when m is not 0.

1. $y^2 - 2xy + x^2 - y + 2x - 1 = 0 \ldots\ldots(1).$

By comparing this with the general equation, Art. (167), we see that

$$a = 1, \quad b = -2, \quad c = 1, \qquad b^2 - 4ac = 4 - 4 = 0;$$

hence, the curve is a parabola.

Making $y = 0$, we obtain

$$x^2 + 2x - 1 = 0; \qquad x = -1 \pm \sqrt{2}.$$

Assuming any line as a unit of measure and laying off

$$AB = -1 + \sqrt{2},$$

$$AB' = -1 - \sqrt{2},$$

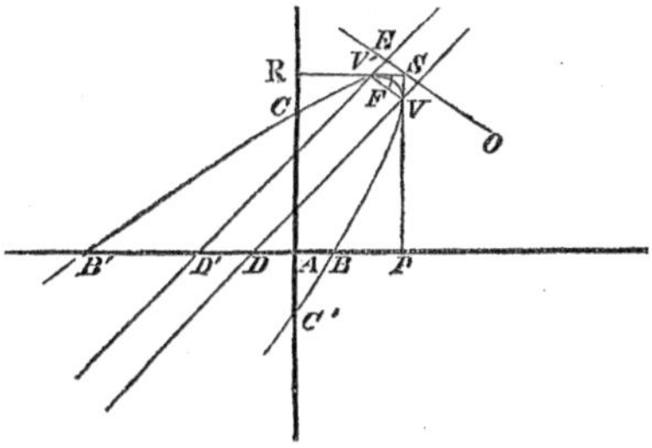

we have the points in which the curve cuts the axis of X. Making $x = 0$, we find

$$y = \frac{1}{2} \pm \sqrt{\frac{5}{4}},$$

and may thus determine the points C and C′ in which the curve cuts the axis of Y.

Solving the given equation, first with reference to y, and then with reference to x, we have

$$y = \frac{2x + 1}{2} \pm \frac{1}{2}\sqrt{-4x + 5} \ldots\ldots (2),$$

$$x = y - 1 \pm \sqrt{-y + 2} \ldots\ldots\ldots\ldots\ldots (3).$$

The equations of the diameters are

$$y = \frac{2x + 1}{2}, \qquad x = y - 1,$$

which represent the lines DV and D′V′.

Placing the quantities under the radical signs (2) and (3), equal to 0, we deduce

$$\text{for the first,} \qquad x = \frac{5}{4};$$

for the second, $y = 2.$

Laying off $AP = \frac{5}{4}$ and drawing the line PV, it must be tangent to the curve at V, and since the coefficient of x under the radical is -4, the curve will lie on the left of this tangent.

Laying off $AR = 2$, and drawing the line RV′, it will be the limit in the direction of the axis of Y, and the curve will be represented as in the figure.

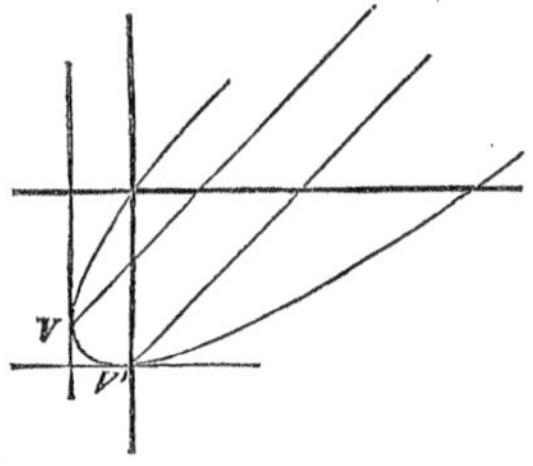

2. $y^2 - 2xy + x^2 + y - 2x = 0.$

3. $y^2 + 2xy + x^2 - 2y - 1 = 0.$

4. $y^2 - 2xy + x^2 - 2y - 2x = 0.$

5. $y^2 + 2xy + x^2 + 2y = 0.$

6. $y^2 - 2xy + x^2 + x = 0.$

Second, when $m = 0$ *and* n *positive.*

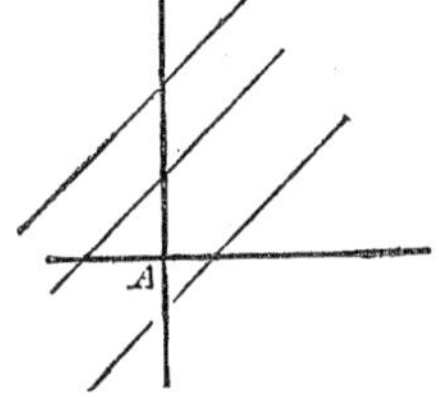

1. $y^2 - 2xy + x^2 - 2y + 2x - 1 = 0.$

2. $y^2 - 2xy + x^2 - 1 = 0.$

3. $y^2 + 4xy + 4x^2 + 4 = 0.$

Third, when $m = 0,\ n = 0.$

1. $y^2 - 2xy + x^2 + 2y - 2x + 1 = 0.$

2. $y^2 - 4xy + 4x^2 = 0.$

Fourth, when $m = 0$, *and* n *negative.*

1. $y^2 + 2xy + x^2 + 1 = 0.$

2. $y^2 + y + 1 = 0.$

174. If it is required to construct the curve with accuracy; we may first solve its equation with reference to y, construct the diameter and determine the limit as in Art. (171). This limit is tangent to the curve at the point in which it intersects the diameter. Solve the equation with reference to x, construct the diameter and determine the limit in the direction of the axis of Y. This is also tangent to the curve at the point in which it intersects the diameter. Since these tangents are parallel to the co-ordinate axes respectively, they are perpendicular to each other and intersect on the directrix, Art. (97). Through their point of intersection draw a line perpendicular to either diameter, it will be the directrix, Art. (100). Join the two points of tangency by a chord, this will pass through the focus, Art. (97). With either point of tangency as a centre, and the distance to the directrix as a radius, describe an arc, it will cut the chord in the focus, Art. (88). Through the focus draw a perpendicular to the directrix, it will be the axis, and the curve may then be constructed as in Art. (88).

To illustrate, let us recur to example (1) in case first, of the preceding article. Having determined the limits PV and RV′, through their point of intersection S, draw SO perpendicular to DV, it is the directrix; join the points V and V′; with V′E describe the arc EF cutting VV′ in F, F is the focus through which the axis may be drawn parallel to DV.

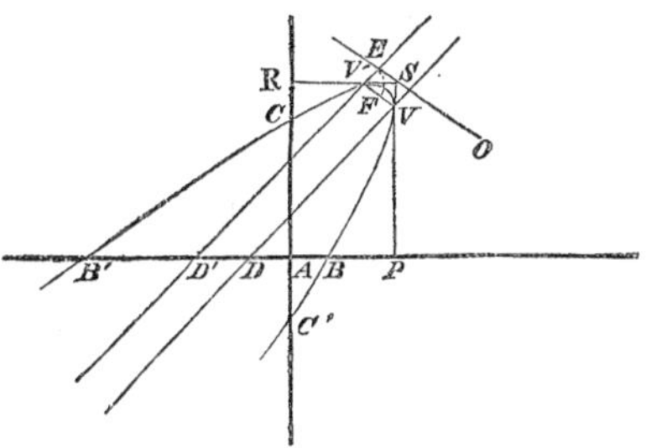

The ellipse when $b^2 - 4ac$ *is negative.*

175. The value of y, equation (1), Art. (168), may be put under the form

$$y = -\frac{bx + d}{2a} \pm \frac{1}{2a}\sqrt{(b^2 - 4ac)\left(x^2 + \frac{2mx}{b^2 - 4ac} + \frac{n}{b^2 - 4ac}\right)}.$$

Those values of x, which will reduce the radical to 0, and give equal values of y, will evidently be obtained, by placing

$$x^2 + \frac{2mx}{b^2 - 4ac} + \frac{n}{b^2 - 4ac} = 0.$$

Solving this equation, and denoting the least value of x by x' and the other by x'', the value of y, may be put under the form

$$y = -\frac{bx + d}{2a} \pm \frac{1}{2a}\sqrt{(b^2 - 4ac)(x - x')(x - x'')}. \quad \ldots.(1).$$

These roots x' and x'' may be *real and unequal*, *real and equal*, or *imaginary*.

When real and unequal. For every value of $x > x''$ the factors $x - x''$ and $x - x'$ will both be positive, their product also positive, and the quantity under the radical sign negative. The corresponding values of y will therefore be imaginary, and there will be no corresponding points of the curve.

For $x = x''$, the quantity under the radical sign is 0, the two values of y equal, and the ordinate produced is tangent to the curve at the vertex of the diameter whose equation is, Art. (168),

$$y = -\frac{bx + d}{2a}.$$

For every value of $x < x''$ and $> x'$, the two factors $x - x''$ and $x - x'$ will have contrary signs, their product will be negative, and the quantity under the radical sign positive, and there will be two corresponding real values of y and two points of the curve.

For $x = x'$, the quantity under the radical sign again becomes 0, and the ordinate will be tangent to the curve at the other vertex of the diameter.

For every value of $x < x'$, the factors $x - x''$ and $x - x'$ will be negative, their product positive, and the values of y imaginary.

Therefore, if two distances AP and AP′, represented by x' and and x'', be laid off on the axis of X, and through their extremities two lines be drawn parallel to the axis of Y, these lines will be tangent to the curve, and no point of the curve can lie without them. Hence, to obtain the limits of the curve in the direction of the axis of abscissas; we solve the equation with reference to y, *place the quantity under the radical sign equal to* 0, *and deduce the roots of the equation, these will be the abscissas of the limits; lay off these abscissas, and through their extremities draw lines parallel to the axis of ordinates, they will be the required limits.* These limits will be tangent to the ellipse at the vertices of the diameter which bisects all chords parallel to the axis of Y.

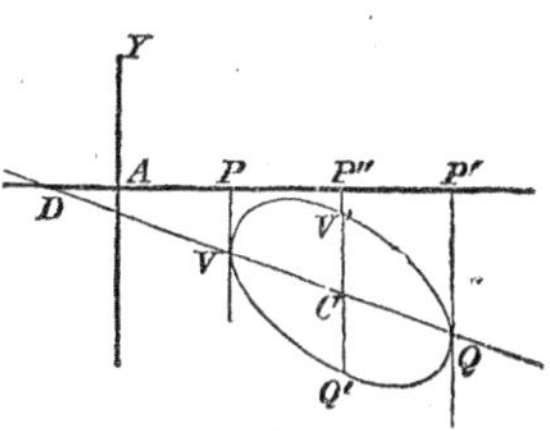

By solving the equation with respect to x, we may obtain, in a similar way, the limits in the direction of the axis of Y.

If the roots x' and x'' are equal, we have

$$(x - x')(x - x'') = (x - x')^2,$$

and the value of y reduces to

$$y = -\frac{bx + d}{2a} \pm \frac{x - x'}{2a}\sqrt{b^2 - 4ac},$$

which will evidently be imaginary for every value of x except $x = x'$, and this gives for the corresponding value of y, denoted by y',

$$y' = -\frac{bx' + d}{2a},$$

y' and x' are then the co-ordinates *of a single point*, to which the ellipse in this case reduces, Art. (168).

If the roots x' and x'' are imaginary, the product $(x - x')(x - x'')$ will be positive for all values of x *; hence, every value of x, in equation (1), will give imaginary values for y, and there can be no points of the curve, which is said in this case *to be imaginary*, Art. (168).

176. An equation of an ellipse being given, the curve may be well represented by following the rule laid down in Art. (173).

Examples.

First, when x' and x'' are real and unequal.

1. $$y^2 - 2xy + 2x^2 + 2y - x = 0,$$

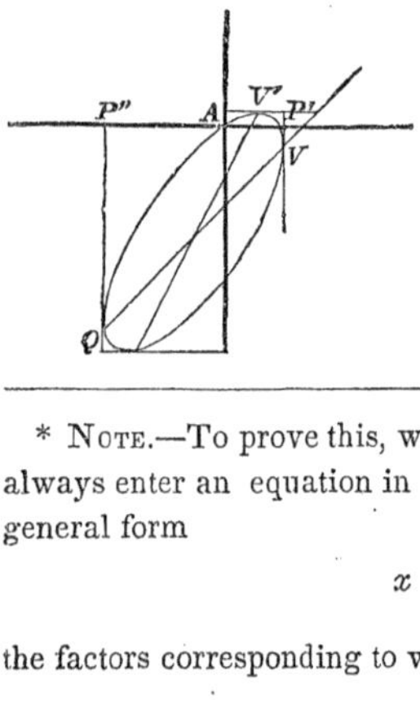

in which $b^2 - 4ac = 4 - 8 = -4$, and

$$y = x - 1 \pm \sqrt{-x^2 - x + 1},$$

$$x' = AP'' = -\frac{1}{2} - \sqrt{\frac{5}{4}},$$

* NOTE.—To prove this, we have only to recollect that imaginary roots always enter an equation in pairs, and must be particular cases of the general form

$$x = a \pm b\sqrt{-1},$$

the factors corresponding to which are

$$x - (a + b\sqrt{-1}) \quad \text{and} \quad x - (a - b\sqrt{-1}),$$

their product being

$$x^2 - 2ax + a^2 + b^2 = (x - a)^2 + b^2,$$

which is evidently positive for all values of x, since it is the sum of two perfect squares.

$$x'' = \mathrm{AP}' = -\frac{1}{2} + \sqrt{\frac{5}{4}}.$$

2. $$y^2 - 2xy + 2x^2 - 2y - 2x = 0.$$

3. $$y^2 + 2xy + 2x^2 - 2x = 0.$$

4. $$2y^2 - 2xy + 3x^2 + 2y + x + 1 = 0.$$

Second, when x' *and* x'' *are real and equal.*

1. $$y^2 - 2xy + 2x^2 - 4x + 4 = 0.$$

2. $$y^2 + x^2 - 2x + 1 = 0.$$

Third, when x' *and* x'' *are imaginary.*

1. $$y^2 + xy + x^2 + x + y + 1 = 0.$$

2. $$y^2 + x^2 + 2x + 2 = 0.$$

177. In order to construct the curve accurately; we solve the equation with reference to y, construct the diameter and determine the abscissas of the limits as in Art. (175). Substituting these in either the equation of the curve or diameter, we find for the ordinates of the vertices V and Q,

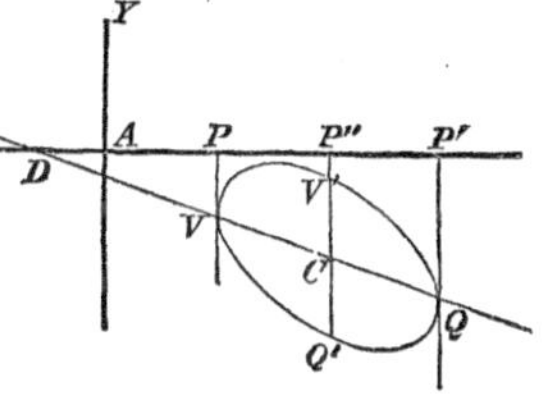

$$y' = -\frac{bx' + d}{2a}, \qquad y'' = -\frac{bx'' + d}{2a.}.$$

Substituting these in expression (2), Art. (17), we have

$$\mathrm{D} = \sqrt{\frac{b^2(x' - x'')^2}{4a^2} + (x' - x'')^2} = \frac{x' - x''}{2a}\sqrt{b^2 + 4a^2} = \mathrm{VQ}.$$

Since this diameter bisects chords parallel to the axis of Y, its

conjugate will be V′Q′, passing through the centre C and parallel to AY, Art. (143). If we denote the abscissa of the point C by z, and substitute it in equation (1), Art. (175), we have for the corresponding values of y, P″V′ and P″Q′,

$$y = -\frac{bz + d}{2a} \pm \frac{1}{2a}\sqrt{(b^2 - 4ac)(z - x')(z - x'')}.$$

The difference of these two values is the length of V′Q′; hence,

$$\text{V}'\text{Q}' = \frac{1}{a}\sqrt{(b^2 - 4ac)(z - x')(z - x'')},$$

or substituting for z its value, which is evidently

$$z = \frac{x' + x''}{2},$$

we have

$$\text{V}'\text{Q}' = \frac{x' - x''}{2a}\sqrt{4ac - b^2}.$$

The length and position of these two conjugate diameters being now known, the curve may be constructed as in Art. (150).

The angle V′CQ, made by the conjugate diameters, may be readily measured, since the tangent of the angle CDP″, in any position of the diameter, will have the same numerical value as tang α, and therefore be equal to $-\dfrac{b}{2a}$ taken with a positive sign; whence, by a reference to a table of natural sines, &c., CDP″ becomes known, and since

$$\text{V}'\text{CV} = 90° - \text{CDP}'',$$

we have

$$\text{V}'\text{CQ} = 180° - \text{V}'\text{CV} = 90° + \text{CDP}''.$$

The two conjugate diameters and the angle made by them

being thus known, the curve may be constructed as in Art. (150), or the axes as well as the angles α and α' may be determined from equations (1), (2), and (3), Art. (157).

The Hyperbola when $b^2 - 4ac$ *is positive.*

178. Let us resume the value of y, equation (1), Art. (175),

$$y = -\frac{bx + d}{2a} \pm \frac{1}{2a}\sqrt{(b^2 - 4ac)(x - x')(x - x'')} \ldots\ldots (1),$$

in which, we must remember that x' and x'' are the values of x obtained by placing the quantity under the radical sign, in the general value of y, equal to zero, and that they will be *real and unequal, real and equal, or imaginary.*

When real and unequal. For every value of $x > x''$, the factors $x - x''$ and $x - x'$ will both be positive, and the quantity under the radical sign positive. The corresponding values of y will therefore be real and unequal, and there will be two corresponding points of the curve.

For $x = x''$ the quantity under the radical sign is zero, and the corresponding ordinate produced will be tangent to the curve at the vertex of that diameter which bisects chords parallel to the axis of Y, Art. (168).

For every value of $x < x''$ and $> x'$, the two factors will have contrary signs, their product will be negative, and the corresponding values of y imaginary, and there will be no corresponding points of the curve.

For $x = x'$, the corresponding ordinate produced, again becomes tangent to the curve at the other vertex of the above diameter.

For every value of $x < x'$, the factors will both be negative, their product positive, and the corresponding values of y real.

Therefore, if two distances AP and AP$'$, represented by x' and x'', be laid off on the axis of X, and through their extremities two

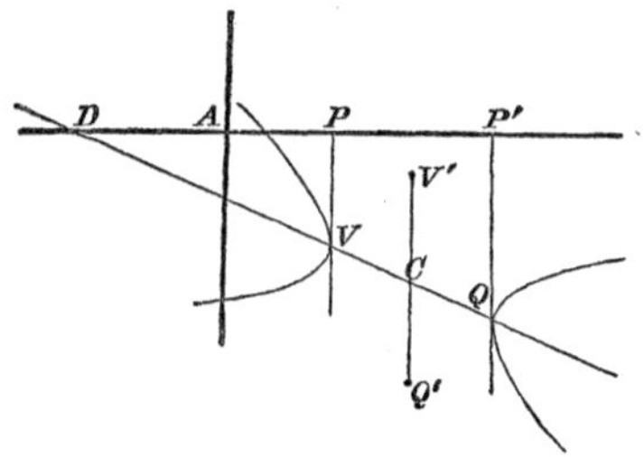

lines be drawn parallel to the axis of Y, these lines will be tangent to the curve, no point of the curve will lie between them, and the curve will extend to infinity in both directions without them. Hence, we obtain the limits of the hyperbola in the direction of either axis of co-ordinates in the same way as described in Art. (175).

If the roots x' *and* x'' *are equal*, the value of y, equation (1), as in the corresponding case in the ellipse, Art. (175), reduces to

$$y = -\frac{bx + d}{2a} \pm \frac{x - x'}{2a}\sqrt{b^2 - 4ac},$$

which will evidently be real for every value of x. This equation then represents *two right lines which intersect*, and to which the hyperbola in this case reduces.

If the roots x' *and* x'' *are imaginary*, the product $(x - x')(x - x'')$ will be positive for all values of x; [see note, Art. (175)], hence every value in equation (1), will give real values for y, and two corresponding points of the curve, and there will be no limits in the direction of the axis of X, as was to be expected, since the abscissas of these limits x' and x'' are imaginary. It also follows, that the diameter which bisects chords parallel to the axis of Y, has no vertices, or does not intersect the curve, which must be as represented in the figure.

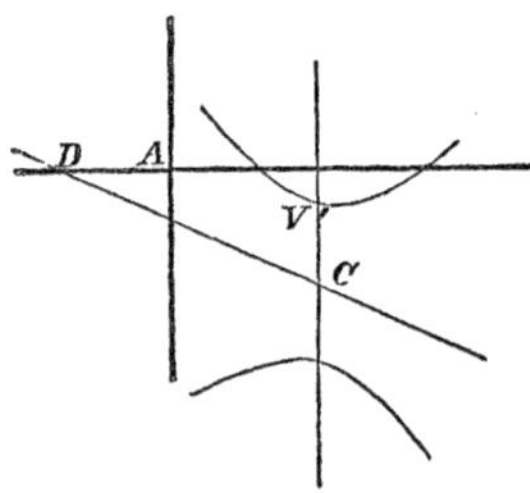

179. An equation of an hyperbola being given, the curve may be well represented by following the rule laid down in Art. (173).

Examples.

First, when x' *and* x'' *are real and unequal.*

1. $y^2 - 2xy - x^2 + 2 = 0.$

in which

$$b^2 - 4ac = 4 - 4 \times 1 \times -1$$
$$= 4 + 4 = 8,$$

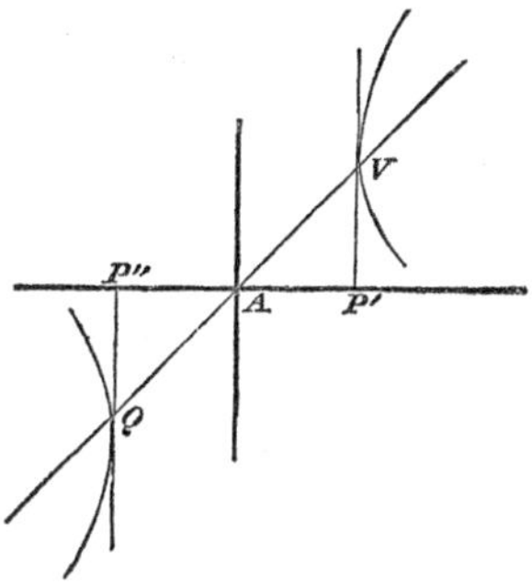

and

$$y = x \pm \sqrt{2x^2 - 2}.$$

2. $y^2 - x^2 + 2x - 2y + 1 = 0.$

3. $y^2 + xy - 2x^2 + x = 0.$

4. $y^2 - 2xy - x^2 - 2y + 2x + 3 = 0.$

Second, when x' *and* x'' *are real and equal.*

1. $y^2 - 2x^2 + 2y + 1 = 0.$

2. $y^2 - x^2 = 0.$

3. $y^2 + xy - 2x^2 + 3x - 1 = 0.$

Third, when x' *and* x'' *are imaginary.*

1. $y^2 - 2xy - x^2 - 2 = 0.$

2. $y^2 + 2xy - x^2 + 2x + 2y - 1 = 0.$

3. $y^2 - 2xy - x^2 - 2x - 2 = 0.$

180. The curve may also be constructed accurately, by first determining the length and position of two conjugate diameters, precisely as in Art. (177). The expressions for these diameters

will be the same as those determined for the ellipse. For the distance cut off by the curve on the one which bisects chords parallel to the axis of Y, we have

$$\frac{x' - x''}{2a}\sqrt{b^2 + 4a^2};$$

and on its conjugate

$$\frac{x' - x''}{2a}\sqrt{4ac - b^2},$$

the first of which will be real, and the second imaginary, when x' and x'' are real, and the reverse when x' and x'' are imaginary.

In this case, the angle V′CD [see figures in Art. (178)], included between the two conjugate diameters, is always equal to 90° − CDA. But we know that tang CDA is numerically equal to $\text{tang}\,\alpha = -\frac{b}{2a}$. We therefore have

$$\text{tang V'CD} = \cot \alpha,$$

from which the angle may at once be found, and then the curve be constructed as in Art. (150), or the axes, together with α and α', may be found from equations (1), (2) and (3) of Art. (158).

OF CENTRES AND DIAMETERS.

181. *The centre of a curve is a point, through which, if any straight line be drawn, terminating in the curve, it will be bisected at this point.*

It follows from this definition, that for each point, as M, of a curve which has a centre, there will be another corresponding point, as M′, on the opposite side of the centre and at the same distance from it. If therefore the origin of co-ordinates be placed at the centre, the co-ordinates of these two points will be equal with contrary signs; that is, if the co-ordinates of one point are $+ x'$ and $+ y'$, those of the other will be $- x'$ and $- y'$, and the equation of the curve must be satisfied by the substitution of each of these sets of co-ordinates. But, this can not be the case, unless all the terms of the equation containing the variables are of an even degree; for if some are of an odd degree, the signs of these terms will be different when $- x'$ and $- y'$ are substituted, from what they are when $+ x'$ and $+ y'$ are substituted, while those of an even degree will remain the same. It is evident then, that the equation can not be satisfied, in both cases.

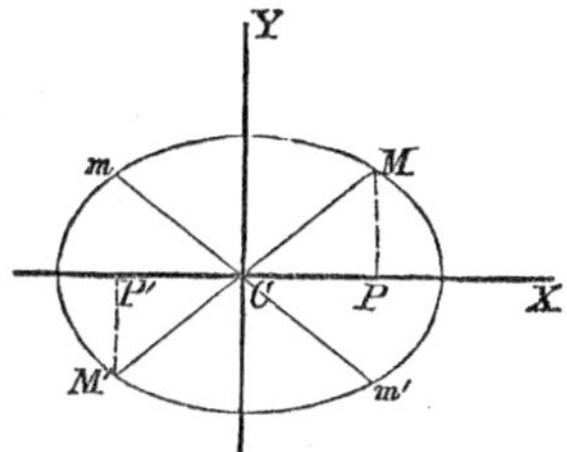

In order then to ascertain whether a given curve has a centre, we first examine its equation and see if all its terms are of an even degree with respect to the variables. If they are, the origin of co-ordinates is a centre. If they are not, we substitute for the variables their values taken from the formulas (2), Art. (67), and see if such values can be assigned to the arbitrary constants a' and b' as will cause all the terms of an odd degree to disappear. If so, the curve will have a centre at the new origin, and the values of a' and b' will be its co-ordinates when referred to the primitive system. If no real and finite values can be thus assigned, the curve will have no centre.

182. To apply the above principles to lines of the second order, we resume the general equation

$$ay^2 + bxy + cx^2 + dy + ex + f = 0,$$

and substitute for x and y their values taken from the formulas of Art. (67).

$$x = a' + x' \qquad y = b' + y'.$$

we thus obtain, after reducing, and denoting the sum of all the terms independent of the variables by f'.

$$ay'^2 + bx'y' + cx'^2 + (2ab' + ba' + d)y' + (2ca' + bb' + e)x' + f' = 0.$$

The terms of this equation will all be of an even degree, if

$$2ab' + ba' + d = 0, \qquad 2ca' + bb' + e = 0,$$

which give for a' and b', the values

$$a' = \frac{2ae - bd}{b^2 - 4ac}, \qquad b' = \frac{2cd - be}{b^2 - 4ac}.$$

These will be real and finite when $b^2 - 4ac$ is not zero, from which we conclude that *there is always a single centre for each ellipse and hyperbola.*

When $b^2 - 4ac = 0$, and the numerators are not zero, the above values reduce to infinity; from which we conclude that, in general, the centre of the parabola is at an infinite distance, or *that the parabola has no centre.*

If $b^2 - 4ac = 0$ and $2ae - bd = 0$, we must also have

$$2cd - be = 0,$$

for by substituting in this the value of $d = \frac{2ae}{b}$, taken from the last expression, it becomes

$$\frac{4ace}{b} - be = 0, \qquad \text{or} \qquad \frac{(4ac - b^2)e}{b} = 0;$$

hence, in this case the two values of a' and b' both become $\frac{0}{0}$, *or indeterminate;* from which we conclude that there are an infinite number of centres, which was plainly to be anticipated, as in this case the parabola reduces to two parallel right lines, Art. (171), and any point of the diameter midway between them will fulfil the condition of a centre.

183. *A diameter of a curve is any straight line which bisects a system of parallel chords drawn in the curve*, Art. (100).

In lines of the second order, if the axis of X be a diameter and the axis of Y be placed parallel to the chords which this diameter bisects, it is evident that the equation of the curve, when referred to these axes, must be of such a form as to give for each single value of x, two values of y, equal with contrary signs. Thus if AX be a diameter, taken as the axis of X, and AY be parallel to the chords which AX bisects, then for each value of x as Ap, the two corresponding values of y, pm and pm', must be equal with contrary signs. This can not be the case as long as the equation of the curve contains any term with the first power of y. The reverse is also true; for if the equation contain no term with the first power of y, for each value of x there will be two equal values of y with contrary signs, and these two values taken together will form a chord bisected by the axis of X. This axis will therefore be a diameter.

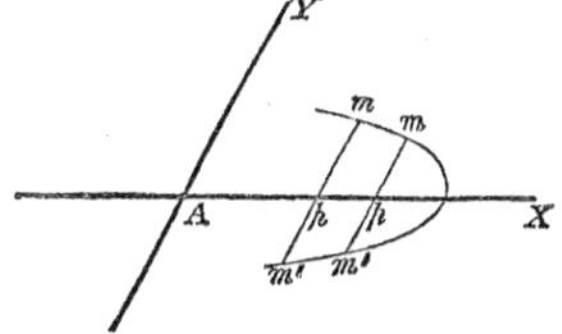

The same reasoning will show that if the axis of Y be a diameter and the axis of X parallel to the chords which it bisects, the equation of the curve can contain no term with the first power of x.

184. Let us now take the general equation of the second degree, Art. (167), and see if by any change of the position of the axes of co-ordinates, we can make either of these axes a diameter. For this purpose, let us substitute for x and y, their values taken

from formulas (3), Art. (67). The new equation, leaving out the dashes of the variables, will be of the form,

$$my^2 + pxy + nx^2 + qy + rx + s = 0,$$

in which

$$m = (a \text{ tang}^2 \alpha' + b \text{ tang } \alpha' + c) \cos^2 \alpha' \ldots\ldots\ldots\ldots(1).$$

$$n = (a \text{ tang}^2 \alpha + b \text{ tang } \alpha + c) \cos^2 \alpha \ldots\ldots\ldots\ldots(2).$$

$$p = (2a \text{ tang } \alpha \text{ tang } \alpha' + b(\text{tang } \alpha + \text{tang } \alpha') + 2c) \cos \alpha \cos \alpha' \ldots(3).$$

$$q = [(2ab' + ba' + d) \text{ tang } \alpha' + (2ca' + bb' + e)] \cos \alpha' \ldots(4).$$

$$r = [(2ab' + ba' + d) \text{ tang } \alpha + (2ca' + bb' + e)] \cos \alpha \ldots(5).$$

If now the axis of X is a diameter, and the axis of Y parallel to the chords which it bisects, we know from the preceding article, that we must have

$$p = 0, \qquad q = 0.$$

We have then to assign such values to the arbitrary quantities α, α', a' and b', as will satisfy the equations

$$2a \text{ tang } \alpha \text{ tang } \alpha' + b(\text{tang } \alpha + \text{tang } \alpha') + 2c = 0 \ldots\ldots\ldots\ldots(6),$$

$$(2ab' + ba' + d) \text{ tang } \alpha' + 2ca' + bb' + e = 0 \ldots\ldots\ldots\ldots(7),$$

and whatever the curve is, this can in general be done; for any value assigned to α in equation (6), taken with the corresponding deduced value of α', will of course satisfy this equation. Tang α' being thus fixed, equation (7) can only be satisfied by means of values attributed to a' and b'. But any value of a' taken with the corresponding deduced value of b' will satisfy this equation.

In the same way it may be shown that if the axis of Y is a diameter, and the axis of X parallel to the chords which it bisects we must have

$$p = 0, \qquad r = 0,$$

and that these equations can always be satisfied.

If both of the axes of co-ordinates are diameters, at the same time, and each parallel to the chords which the other bisects, we must have

$$p = 0, \qquad q = 0, \qquad r = 0.$$

We have seen above, that it is always possible to satisfy the equation $p = 0$,......(6), by assigning at pleasure a value to either α or α', and deducing the corresponding value of the other. These two angles being determined, a proper direction is given to the new axes of co-ordinates, while the new origin is yet to be fixed, so that we may have at the same time

$$q = 0, \qquad r = 0;$$

that is

$$(2ab' + ba' + d) \text{ tang } \alpha' + (2ca' + bb' + e) = 0,$$

$$(2ab' + ba' + d) \text{ tang } \alpha + (2ca' + bb' + e) = 0.$$

These equations being the same, except that tang α in one, occupies the place of tang α' in the other, it is evident they can not both be satisfied, at the same time, unless we have the terms separately equal to 0, that is,

$$2ab' + ba' + d = 0, \qquad 2ca' + bb' + e = 0,$$

which give for a' and b' the values

$$a' = \frac{2ae - bd}{b^2 - 4ac}, \qquad b' = \frac{2cd - be}{b^2 - 4ac}.$$

We recognise these values as the co-ordinates of the centre of the curve, Art. (182), and therefore conclude that the new origin must be at the centre, and that the new axes are conjugate diameters, Art. (143). And since the above values are finite only for the ellipse and hyperbola, and infinite for the parabola, we conclude that *both of the co-ordinate axes may be diameters at the same time in the ellipse and hyperbola, but not in the parabola.*

And since there are an infinite number of values of α and a which will fulfil the above conditions, we conclude that *in the ellipse and hyperbola, there is an infinite number of conjugate diameters.*

We have seen above that equation $p = 0$, being satisfied, the axis of X will be a diameter, if we also have

$$q = 0.$$

If in this equation (7) we regard a' and b' as variables, it will be the equation of a straight line, and any values of a' and b' which are the co-ordinates of a point on this line will satisfy the equation, Art. (23); hence, the new origin may be any where on this line. But this new origin *must be* on the new axis of X, and *may be* any where on this axis, (now a diameter of the curve). Hence

$$q = 0,$$

must be the equation of this new axis of X, or diameter, referred to the primitive axes, a' and b' being the variables.

If the axis of Y be made a diameter, similar reasoning will show that $r = 0$ will be the equation of this diameter.

The fact that $q = 0$ is the equation of a diameter, leads to two important conclusions.

First. Since by assigning all possible values to α' this equation may be made to represent all possible diameters, and since the co-ordinates of the centre, Art. (182), when substituted for a' and b', in this equation, must satisfy it, as they were obtained by placing

$$2ab' + ba' + d = 0, \qquad 2ca' + bb' + e = 0,$$

we conclude that *every diameter passes through the centre.*

Second. If any straight line be drawn through the centre, and the origin of co-ordinates be placed at the centre, and the right line be taken as the axis of X, the values of a' and b' will satisfy the equation $q = 0$; and the position of the line being given, α is known, and the corresponding value of α', deduced from the equation $p = 0$, will satisfy it also and give a proper direction to the axis of Y. Both of these equations being thus satisfied, we conclude that the right line is a diameter; hence, *every right line passing through the centre is a diameter.*

185. We have seen in the preceding article, that both axes of co-ordinates can not be diameters in the parabola, but that the axis of X will be a diameter and the axis of Y parallel to the chords which it bisects, when

$$p = 0, \qquad q = 0,$$

and as the equation when referred to these axes is still the equation of the parabola, we must have, Art. (169),

$$p^2 - 4mn = 0,$$

and since $p = 0$, $-4mn$ must equal 0. But m can not be 0, for if it were, the equation referred to the new axes would reduce to

$$nx^2 + rx + s = 0,$$

which is the equation of no curve; hence, *we must have* $n = 0$, and the equation will reduce to

$$my^2 + rx + s = 0.$$

Hence, in the parabola $n = 0$ is a condition consequent upon $p = 0$ and $q = 0$.

This fact may be verified thus: Since in the parabola all diameters are parallel, and make with the axis of X an angle whose tangent is $-\frac{b}{2a}$, Art. (172), and since the new axis of X is a diameter, we have

$$\text{tang}\, \alpha = -\frac{b}{2a}.$$

Substituting this value in equation (2), Art. (184), we have

$$\frac{n}{\cos^2 \alpha} = \frac{ab^2}{4a^2} - \frac{b^2}{2a} + c = -\frac{b^2}{4a} + c = -\frac{b^2 - 4ac}{4a} = 0.$$

If the axis of Y is a diameter, it may be proved, in the same way, that we must have $m = 0$, and that the equation of the parabola will take the form

$$nx^2 + qy + s = 0.$$

It may be further remarked, that any value whatever being assumed either for *tang* α or *tang* α' and substituted in equation (6), will, *for the parabola*, give $-\frac{b}{2a}$ for the value of the other. Also, if $-\frac{b}{2a}$ be substituted in the same equation for tang α or tang α', the corresponding value of the other will be $\frac{0}{0}$, or indeterminate. This is evidently a consequence of the parallelism of the diameters of the parabola.

OF LOCI.

186. The term *locus*, in Analytical Geometry is applied to the line or surface, in which are to be found all of the positions of a point or line, which changes its position in accordance with some determinate law.

Thus, if a point is moved in a plane, so that it shall always be at the same distance from a fixed point, *the locus* of the point will be the circumference of a circle.

Also, a plane tangent to a surface at a given point, is *the locus* of all right lines drawn tangent to lines of the surface at this point.

187. The determination of *the loci* of points, which are moved in a given plane subject to certain conditions, gives rise to a great variety of interesting problems, several of which it is proposed to solve and discuss in detail, for the purpose of indicating to the student the general method to be pursued in the solution of all.

It should be remarked, that pains should be taken to select the best position for the co-ordinate axes in each problem, as its solution may be thus much simplified.

188. *Problem* 1*st.* To determine the locus of a point, which in any of its positions is at equal distances from a fixed point and fixed right line.

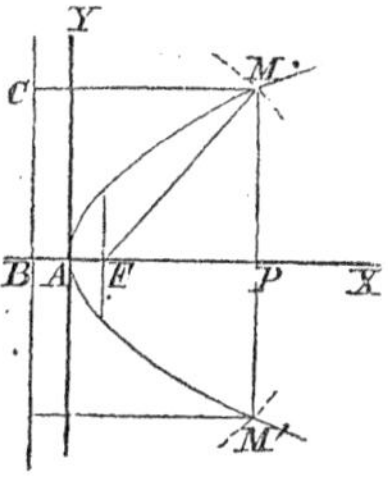

Let F be the given point and BC the given right line. Through F draw FB perpendicular to BC and denote the known distance FB by p. At the middle point of FB erect AY perpendicular to it and take AX and AY as the co-ordinate axes. Let M be any position of the moving point, the co-ordinates of which are AP $= x$, and PM $= y$. By the conditions of the problem, we must have

$$\text{MF} = \text{MC}.$$

But

$$\text{MF} = \sqrt{\overline{\text{MP}}^2 + \overline{\text{FP}}^2} = \sqrt{y^2 + \left(x - \frac{p}{2}\right)^2},$$

and

$$\text{MC} = \text{BP} = \text{BA} + \text{AP} = x + \frac{p}{2};$$

hence

$$\sqrt{y^2 + \left(x - \frac{p}{2}\right)^2} = x + \frac{p}{2}.$$

Squaring both members and reducing, we obtain

$$y^2 = 2px,$$

an equation expressing the relation between x and y for all positions of the point M. It is therefore the equation of the locus, which is a parabola, Art. (88).

189. *Problem 2nd.* To find the locus of a point moving in such a way, that the sum of its distances from two given points shall always be equal to a given line.

Let F and F′ be the two given points, and $2c$ the distance between them. Let $2a$ represent the given line.

At C, the middle point of FF′, erect the perpendicular CD and take CF and CD as the co-ordinate axes. Let M be any position of the point and denote its co-ordinates by x and y, and denote by r and r' the distances from the point to F and F′.

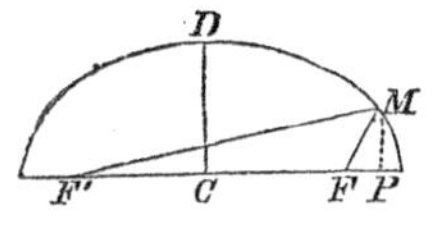

The right angled triangle FMP gives

$$\overline{FM}^2 = \overline{MP}^2 + \overline{FP}^2,$$

or, since $CF = c$,

$$r^2 = y^2 + (x - c)^2.$$

In the same way the right angled triangle F′MP, gives

$$r'^2 = y^2 + (x + c)^2.$$

Adding these two equations, member by member, we have

$$r^2 + r'^2 = 2(y^2 + x^2 + c^2) \ldots\ldots\ldots\ldots (1),$$

and subtracting them,

$$r'^2 - r^2 = 4cx, \quad \text{or} \quad (r + r')(r - r') = 4cx\ldots\ldots(2).$$

But by the condition of the problem,

$$r' + r = 2a\ldots\ldots\ldots\ldots(3).$$

Substituting this in equation (2), we have

$$r' - r = \frac{2cx}{a}$$

Combining this with (3), we deduce

$$r' = a + \frac{cx}{a}, \qquad r = a - \frac{cx}{a}\ldots\ldots\ldots\ldots(4).$$

Squaring these values and substituting in (1), we obtain

$$a^2 + \frac{c^2x^2}{a^2} = y^2 + x^2 + c^2,$$

or

$$a^2y^2 + (a^2 - c^2)\,x^2 = a^2\,(a^2 - c^2),$$

or putting b^2 for $a^2 - c^2$,

$$a^2y^2 + b^2x^2 = a^2b^2,$$

the same as equation (e), Art. (105), and the locus is an ellipse.

190. *Problem 3d.* To find the locus of any point of a given right line, which is moved so that its extremities shall be constantly in two other right lines, at right angles to each other.

Let AX and AY be the two right lines at right angles, and M any point of the given line CB. Denote the distance BM by a, and MC by b. Take AX and AY as the co-ordinate axes and let AP $= x$, PM $= y$.

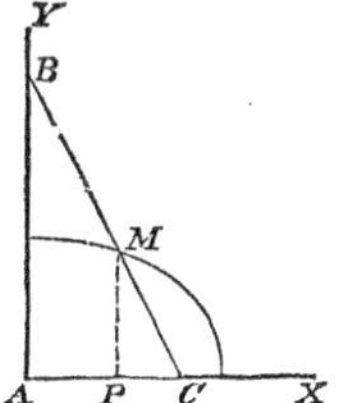

Since MP is parallel to AB, we have

$$\text{PC} : \text{MC} :: \text{AP} : \text{BM},$$

or

$$\sqrt{b^2 - y^2} : b :: x : a;$$

whence

$$b^2x^2 = a^2b^2 - a^2y^2, \quad \text{or} \quad a^2y^2 + b^2x^2 = a^2b^2,$$

which is evidently the equation of an ellipse whose semi-transverse axis is BM and semi-conjugate MC.

191. *Problem 4th.* To find the locus of the centres of all circles which pass through a given point and are tangent to a given right line.

Let M be the given point, and BX the given line. Through M, draw MA perpendicular to BX, and let AX and AM be the axes of co-ordinates. Denote the ordinate MA by p, the abscissa of this point will be 0. Let C be the centre of one of the circles and denote its co-ordinates by x' and y'. The equation of this circle, Art. (34), will be

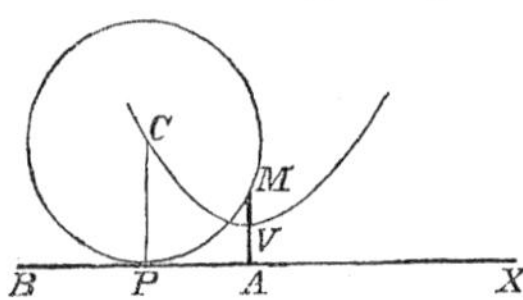

$$(x - x')^2 + (y - y')^2 = R^2.$$

But since it passes through the point M, the co-ordinates of this point will satisfy the equation, and give

$$x'^2 + (p - y')^2 = R^2,$$

and since the circle is tangent to BX, we have $R = y'$; hence

$$x'^2 + (p - y')^2 = y'^2,$$

or

$$x'^2 - 2py' = -p^2,$$

which expresses the relation between x' and y' for any position of the circle, it is therefore the equation of the locus.

If the origin be now transferred to V midway between M and A the formulas (2) of Art. (67) become

$$x' = x, \qquad y' = y + \frac{p}{2},$$

the substitution of which gives

$$x^2 = 2py,$$

the equation of a parabola of which M is the focus and BX the directrix, and this is evidently another method of enunciating and solving problem 1st, Art. (188).

192. *Problem 5th.* To find the locus of the intersection of right lines, drawn from the extremities of the transverse axis of a given ellipse, to the extremities of chords of the ellipse perpendicular to the transverse axis.

Let ABD be the given ellipse and DD′ any chord perpendicular to AB. Through D and D′ draw the lines AD and BD′, it is required to find the locus of M, their point of intersection. Let the equation of the given ellipse be

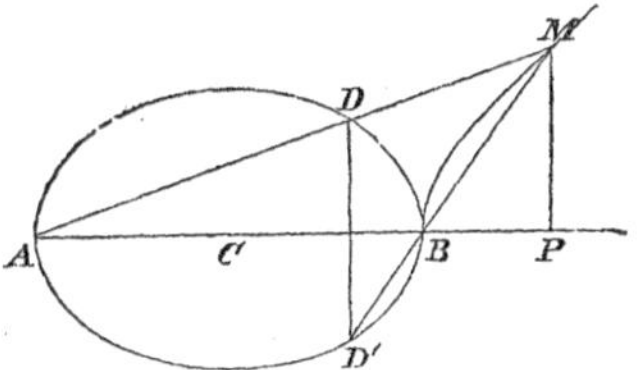

$$a^2y^2 + b^2x^2 = a^2b^2,$$

and denote the co-ordinates of the point D by x' and y'. The equation of condition that this point shall be on the ellipse will be

$$a^2y'^2 + b^2x'^2 = a^2b^2, \quad \text{or} \quad y'^2 = \frac{b^2}{a^2}(a^2 - x'^2) \ldots\ldots(1).$$

The equation of the right line AD, passing through the two points A and D, Art. (31), will be

$$y = \frac{y'}{x' + a}(x + a) \ldots\ldots\ldots\ldots (2),$$

and of the line D′B,

$$y = \frac{-y'}{x' - a}(x - a) \ldots\ldots\ldots\ldots (3).$$

Multiplying these equations, member by member, we have

$$y^2 = \frac{-y'^2}{x'^2 - a^2}(x^2 - a^2) \ldots\ldots (4),$$

in which y and x are the co-ordinates of the point of intersection, for the two particular lines AD and D′B. If y' and x' be eliminated from this equation, it is evident that y and x will belong to no particular lines, but will be the co-ordinates of the point of intersection of all the lines which fulfil the required condition; and the resulting equation will be the equation of the required locus. Substituting the value of y'^2 taken from equation (1) in equation (4), it reduces to

$$y^2 = \frac{b^2}{a^2}(x^2 - a^2),$$

which is the equation of an hyperbola having the same axes as the given ellipse, Art. (105).

This method of determining loci, by combining two equations belonging to particular lines, so as to eliminate the arbitrary constants which serve to determine the position of the lines, thus deducing an equation independent of these constants, and therefore belonging to all lines which fulfil the required condition, is of frequent use.

193. *Problem 6th.* If from the extremity of a diameter of a circle

any straight line, as AR, be drawn until it intersects the tangent BR at the other extremity, and the distance AM be laid off equal to NR, it is required to find the locus of M. Let A be the origin, and AB and AY the co-ordinate axes. Let AB $= 2a$, AP $= x$, PM $= y$. Then drawing NP′ parallel to MP, we have

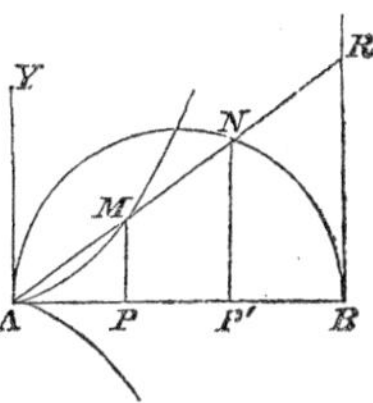

$$\text{AP} : \text{PM} :: \text{AP}' : \text{P}'\text{N}.$$

Also, since AM = NR, AP = P′B,

$$\text{P}'\text{N} = \sqrt{\text{P}'\text{B} \times \text{AP}'} = \sqrt{x(2a - x)}.$$

The above proportion then becomes

$$x : y :: 2a - x : \sqrt{x(2a - x)};$$

whence

$$y^2 = \frac{x^3}{2a - x} \qquad \text{or} \qquad y = \pm\sqrt{\frac{x^3}{2a - x}};$$

for the equation of the locus. The equation being of the third degree, the line is of the third order, Art. (33).

All negative values of x give imaginary values for y.

$x = 0$ gives $y = \pm 0$.

Each positive value of $x < 2a$ gives two real values of y, equal with contrary signs.

$x = 2a$ gives $y = \pm \infty$.

All positive values of $x > 2a$ give imaginary values for y, and the curve is as indicated in the figure, the line BR being an asymptote, Art. (161). It is called the *Cissoid of Diocles.*

194. The following problems may be solved by pursuing methods similar to those indicated in the preceding articles.

7. To find the locus of a point moving in such a way, that the difference of its distances from two given points shall always be equal to a given line.

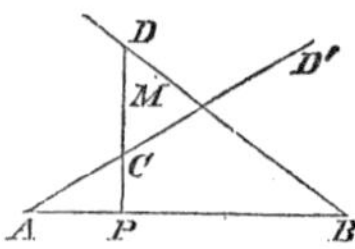

8. Given the line AB and the two lines DB and AD′, to find the locus of M moving so that MP shall be a mean proportional between PC and PD.

9. Given the base of a triangle and the difference of the angles at the base, to find the locus of the vertex.

10. Given the base of a triangle, to find the locus of the vertex when one angle at the base is double of the other.

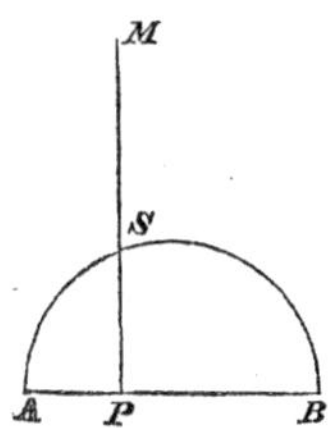

11. To find the locus of the point of intersection of a tangent to an ellipse, with a perpendicular let fall upon it from either focus.

12. Given the semi-circle ASB, to find the locus of the point M, so that we may always have

$$AP : PS :: AB : PM.$$

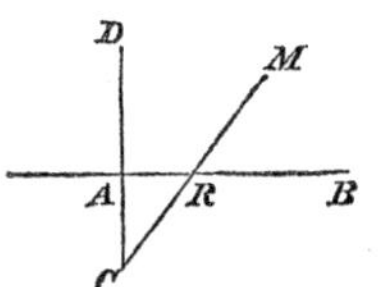

13. Given the indefinite right line AB, the point C, and the perpendicular CD, to find the locus of M so that we may always have $MR = AD$.

OF SURFACES OF REVOLUTION.

195. *A surface of revolution, is a surface which may be generated by revolving a line about a right line as an axis.*

By *revolving*, is to be understood, moving the line in such a manner, that each point of it will generate the circumference of a circle whose centre is in the axis, and whose plane is perpendicular to the axis. The moving line is called *the generatrix.*

From the definition it follows, that every plane perpendicular to the axis will cut a circle from the surface.

Every plane passed through the axis will cut from the surface *a meridian curve, or line*, and if this be revolved about the axis it will generate the surface.

196. In order to obtain the general equation of a surface of revolution, Art. (54), let us take the axis of the surface for the axis of Z, and the co-ordinate planes at right angles. The general equation of the generatrix will then be, Art. (52),

$$x = f(z), \qquad y = f'(z) \ldots\ldots\ldots\ldots (1),$$

and let r denote the distance of any point of this line from the axis. Since, from the nature of the surface, this point in its revolution must describe a circle whose centre is in the axis of Z, and whose plane is perpendicular to this axis, that is parallel to the plane XY, we must have in every position of the point,

$$x^2 + y^2 = r^2 \ldots\ldots\ldots\ldots (2),$$

and since this point is on the generatrix, the values of x and y taken from equations (1), must fulfil the condition expressed by equation (2), and give

$$\overline{f(z)}^2 + \overline{f'(z)}^2 = r^2.$$

Equating these two values of r^2, we have

$$x^2 + y^2 = \overline{f(z)}^2 + \overline{f'(z)}^2 \ldots\ldots\ldots\ldots (3),$$

an equation expressing the relation between the co-ordinates of the point in all of its positions. It is therefore *the equation of the surface*, in which $f(z)$ and $f'(z)$, are the values of x and y obtained by solving the equations of the generatrix.

197. To illustrate, let us find the equation of a surface generated by revolving a right line about an axis not in the same plane with it.

The axis of revolution being taken as the axis of Z, we may take for the equations of the generatrix, Art. (44),

$$x = az + \alpha, \qquad y = bz + \beta,$$

from which, we have

$$f(z) = az + \alpha, \qquad f'(z) = bz + \beta.$$

Substituting these in equation (3), it becomes

$$x^2 + y^2 = (az + \alpha)^2 + (bz + \beta)^2.$$

If the axis of X be assumed perpendicular to the generatrix and intersecting it, the projection of the generatrix on the plane XZ will be parallel to the axis of Z, and its projection on the plane YZ will pass through the origin of co-ordinates ; hence, Art. (45), we have

$$a = 0, \qquad \beta = 0,$$

and the above equation becomes

$$x^2 + y^2 - b^2z^2 = \alpha^2 \ldots\ldots\ldots\ldots (1).$$

If we intersect this surface by a plane parallel to XY, the equation of which, Art. (62), is

$$z = c, \qquad x \text{ and } y \quad \textit{indeterminate},$$

we shall obtain, Art. (62),

$$x^2 + y^2 = b^2c^2 + \alpha^2,$$

for the equation of the projection of the intersection on the plane XY, which represents a circle whose radius is $\sqrt{b^2c^2 + \alpha^2}$, Art. (35) ; and this circle will be real, whatever be the value of c ; and the smallest possible when $c = 0$, in which case the cutting plane is the plane XY, Art. (62). And since this projection is equal to the intersection itself, we see that every intersection by a plane perpendicular to the axis will be a circle, as we know it should be, from the definition of the surface.

If we make $y = 0$ in equation (1), we have

$$x^2 - b^2z^2 = a^2, \qquad \text{or} \qquad b^2z^2 - x^2 = -a^2,$$

for the intersection by the plane XZ, Art. (62).

If we make $x = 0$, we have for the intersection by the plane YZ,

$$b^2z^2 - y^2 = -a^2,$$

and these are evidently the equations of two equal hyperbolas, the conjugate axis of each lying on the axis of Z, Art. (105). And since the surface may be generated by revolving either of these meridian curves about the axis, it is called *a hyperboloid of revolution of one nappe.* Of *one* nappe, since, as is readily seen, it forms one uninterrupted surface.

198. If the generatrix is in the plane with the axis of revolution, this plane may be taken for the plane XZ, and as before, the axis of revolution for the axis of Z, in which case the equations of the generatrix will be, Art. (52),

$$x = f(z), \qquad y = f'(z) = 0,$$

and equation (3) of Art. (196) will reduce to

$$x^2 + y^2 = \overline{f(z)}^2 \ldots\ldots\ldots\ldots (1),$$

in which $f(z)$ is the value of x deduced from the equation of the generatrix.

Examples.

1. The equation of a right line in the plane XZ, and passing

through a point on the axis of Z, whose co-ordinates are $x' = 0$, $z' = c$, will be, Art. (29),

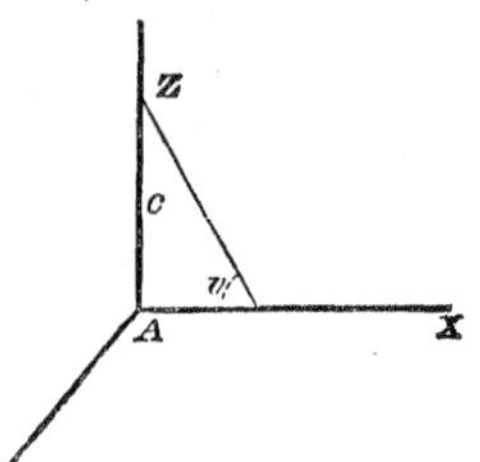

$$x = a(z - c),$$

from which we have

$$f(z) = a(z - c).$$

This substituted in equation (1), gives

$$x^2 + y^2 = a^2(z - c)^2,$$

for the equation of the cone generated by revolving the right line about the axis of Z. This equation may be put under the form

$$(x^2 + y^2)\frac{1}{a^2} = (z - c)^2,$$

or denoting the angle made by the generatrix with the base by v, we have

$$\frac{1}{a} = \text{tang } v;$$

whence

$$(x^2 + y^2) \text{ tang}^2 v = (z - c)^2,$$

the same equation as that deduced in Art. (80).

2. If the axis of a parabola in the plane XZ, coincide with the axis of Z, and its vertex be at the origin of co-ordinates, its equation will be, Art. (84),

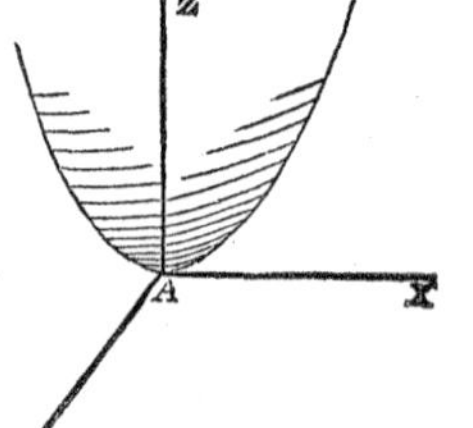

$$x^2 = 2pz,$$

from which we have

$$f(z) = \sqrt{2pz}, \qquad \overline{f(z)}^2 = 2pz,$$

which substituted in equation (1), gives

$$x^2 + y^2 = 2pz,$$

for the equation of the surface generated by revolving a parabola about its axis; called *a paraboloid of revolution.*

3. If the transverse axis of an ellipse, in the plane XZ, lies on the axis of Z, and its centre is at the origin of co-ordinates, its equation will be, Art. (105),

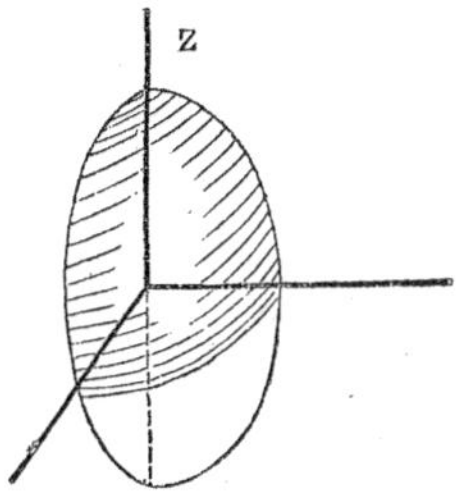

$$a^2x^2 + b^2z^2 = a^2b^2,$$

whence

$$x^2 = \frac{b^2}{a^2}(a^2 - z^2) = \overline{f(z)}^2,$$

and this in equation (1), gives

$$x^2 + y^2 = \frac{b^2}{a^2}(a^2 - z^2), \quad \text{or} \quad a^2(x^2 + y^2) + b^2z^2 = a^2b^2 \ldots (2),$$

for the equation of a surface generated by revolving an ellipse about its transverse axis.

If the conjugate axis of the ellipse lies on the axis of Z, the equation will be,

$$a^2z^2 + b^2x^2 = a^2b^2, \quad \text{whence} \quad x^2 = \frac{a^2}{b^2}(b^2 - z^2) = \overline{fz}^2,$$

and the equation of the surface

$$b^2(x^2 + y^2) + a^2z^2 = a^2b^2 \ldots\ldots\ldots\ldots (3).$$

These surfaces are called *ellipsoids of revolution*; or *spheroids.* The first is *the prolate*, and the second *the oblate spheroid.*

If in either of equations (2) and (3) we make $a = b$, the ellipse becomes a circle, and the equation reduces to

$$x^2 + y^2 + z^2 = a^2,$$

for *the equation of a sphere.*

4. If in equations (2) and (3) we change b^2 into $-b^2$, we have

$$a^2(x^2 + y^2) - b^2z^2 = - a^2b^2,$$

and

$$b^2(x^2 + y^2) - a^2z^2 = a^2b^2.$$

The first represents the surface generated by revolving an hyperbola about its transverse axis, or *hyperboloid of revolution of two nappes.* Of two nappes, since it consists of two distinct parts, one being generated by one branch of the hyperbola, and the other by the other branch.

The second represents the surface generated by revolving the hyperbola about its conjugate axis. Its equation, after dividing by b^2, becomes

$$x^2 + y^2 - \frac{a^2}{b^2}z^2 = a^2,$$

of the same form as equation (1), Art. (197). From which we see that this surface may not only be generated by revolving an hyperbola about its conjugate axis, but also by revolving a right line about another, not in the same plane with it.

OF SURFACES OF THE SECOND ORDER.

199. Surfaces, like lines, Art. (33), are classed into orders according to the degree of their equations.

We have seen, Art. (57), that the plane is the only surface of the first order.

The equation of every surface of the second order must be a particular case of the most general equation of the second degree between three variables,

$$mx^2 + ny^2 + pz^2 + m'xy + n'xz + p'yz$$
$$+ m''x + n''y + p''z + l = 0 \text{..........} (1),$$

which, for the same reason as that given in Art. (167), may be considered as referred to a system of co-ordinate planes at right angles.

Points of the surfaces may be determined as in Art. (55), by assigning values to x and y, and deducing the corresponding values of z; but the nature of the surface will, in general, be best ascertained by intersecting it by planes and discussing the curves of intersection thus obtained.

200. If we combine the above equation, with the equation of a plane having any position, Art. (55), and then refer the line of intersection to co-ordinate axes in its own plane, the resulting equation will be of the second degree. For one of the equations being of the first, and the other of the second degree, the result of their combination will necessarily be of the second degree. We therefore conclude, that the line of intersection of any surface of the second order by a plane, *is a line of the second order, or one of the conic sections*, Art. (170).

201. In the surface represented by the general equation of Art. (199), conceive a system of parallel chords to be drawn. The equations of one of these chords will be of the form, Art. (44),

$$x = az + \alpha, \qquad y = bz + \beta \ldots\ldots\ldots\ldots(1),$$

and these equations may be made to represent any chord of the system, by giving proper values to α and β, a and b remaining unchanged. If equations (1) be combined with the general equation (1), Art. (199), and x and y be eliminated, a result will be obtained of the form

$$z^2 + \frac{s}{r}z + \frac{t}{r} = 0,$$

in which the two values of z will be the ordinates of the points in which the chord pierces the surface. If x', y' and z' denote the co-ordinates of the middle point of this chord, since z' will equal the half sum of the two values of z, we shall have

$$z' = -\frac{s}{2r},$$

or putting for s and r their values, as found by the actual combination of the equations,

$$z' = -\frac{\alpha(2ma + m'b + n') + \beta(2nb + m'a + p') + m''a + n''b + p''}{2(ma^2 + nb^2 + p + m'ab + n'a + p'b)}.$$

Since the point (x', y', z') is on the chord, we also have

$$x' = az' + \alpha, \qquad y' = bz' + \beta.$$

If now these three equations be combined, so as to eliminate α and β; x', y' and z' will belong to the middle point of no particular chord, and the resulting equation will therefore represent the locus of the middle points of all the chords of the system, Art. (192).

Combining the equations, by substituting for α and β, in the first, their values taken from the last, we obtain after reduction,

$$z' = -\frac{(2ma + m'b + n')x' + (2nb + m'a + p')y' + m''a + n''b + p''}{2p + n'a + p'b},$$

which is *the equation of a plane*, Art. (57). We therefore conclude, *that every system of parallel chords of the surface may be bisected by a plane.*

In order that this plane shall be perpendicular to the chords which it bisects, we must have the two conditions, Art. (59),

$$a = \frac{2ma + m'b + n'}{2p + n'a + p'b}, \qquad b = \frac{2nb + m'a + p'}{2p + n'a + p'b},$$

and these equations can always be satisfied by at least one set of real values for a and b; for if they be combined and either a or b

eliminated, there will result an equation of the third degree, containing the other, which must have at least *one real root*, and may have three. Hence, *in every surface of the second order, there is at least one plane which is perpendicular to the system of chords which it bisects.*

202. Let such plane be taken as the co-ordinate plane XY, the axis of Z being perpendicular to it, that is, parallel to the chords. This plane will intersect the surface in a line of the second order, Art. (200), the axis of which may be determined as in Art. (100) or (154). Let this axis be taken as the axis of X and a line, perpendicular to it in the plane XY, as the axis of Y, and suppose the surface to be referred to this new system of co-ordinate planes.

Since the plane XY bisects a system of chords parallel to the axis of Z, the equation of the surface must be of such a form, that for every value of x and y, it must give two equal values of z with contrary signs. It can therefore contain no term involving the first power of z, Art. (183). We must then have in the general equation of Art. (199),

$$n' = 0, \qquad p' = 0, \qquad p'' = 0 \ldots\ldots\ldots\ldots (1).$$

And since the axis of X bisects all chords in the plane XY, parallel to the axis of Y, the equation of the surface must also be of such a form that for all values of x, (z being equal to 0), there must be two equal values of y with contrary signs. The equation can then contain no term involving the first power of y. We must therefore have, in addition to the above equations (1),

$$m' = 0, \qquad n'' = 0,$$

and the general equation (1), Art. (199), must reduce to the form

$$mx^2 + ny^2 + pz^2 + m''x + l = 0 \ldots\ldots\ldots\ldots (3);$$

and as the above transformations are always possible, this equation may be made to represent all surfaces of the second order by assigning proper values to the constants which enter it.

203. To discuss the above equation more fully, let us first transfer the origin of co-ordinates to a point on the axis of X, at a distance from the primitive origin represented by the arbitrary quantity a', the axes remaining parallel to the primitive. The formulas of Art. (72) become

$$x = a' + x', \qquad y = y', \qquad z = z'.$$

Substituting in the above equation, we obtain

$$mx'^2 + ny'^2 + pz'^2 + (2ma' + m'')\,x' + ma'^2 + m''a' + l = 0 \ldots (1).$$

Since a' is arbitrary, we may assign to it such a value as to make

$$2ma' + m'' = 0, \qquad \text{or} \qquad a' = -\frac{m''}{2m},$$

in which case the equation, after denoting the absolute term by l' and omitting the dashes of the variables, reduces to

$$mx^2 + ny^2 + pz^2 + l' = 0 \ldots\ldots\ldots\ldots (2).$$

If $m = 0$, this transformation will in general be impossible, as we shall then have

$$a' = -\frac{m''}{0} = \infty \ldots\ldots\ldots\ldots (3).$$

In this case we may assign to a' such a value as will make

$$m''a' + l = 0, \qquad \text{or} \qquad a' = -\frac{l}{m''},$$

and equation (1) will reduce to

$$ny^2 + pz^2 + m''x = 0 \ldots\ldots\ldots (4).$$

If, however, we have at the same time $m'' = 0$, this transformation will be impossible. But in this case, equation (1) will at once reduce to

$$ny^2 + pz^2 + l = 0, \qquad x \textit{ indeterminate} \ldots\ldots\ldots\ldots (5),$$

which is evidently the equation of a right cylinder with an elliptical or hyperbolic base, according as n and p have the same or contrary signs, Art. (170), the axis of the cylinder coinciding with the axis of X. Moreover, in this case equation (3) gives

$$a' = \frac{0}{0} \quad \textit{indeterminate},$$

and any point of the axis of X will fulfil the required condition.

If $m = 0$, $n = 0$, equation (3), Art. (202), reduces to

$$pz^2 + m''x + l = 0, \qquad y \textit{ indeterminate}.$$

If $m = 0$, $p = 0$, it reduces to

$$ny^2 + m''x + l = 0, \qquad z \textit{ indeterminate};$$

both of which are equations of right cylinders with parabolic bases, the axis of the first being parallel to the axis of Y, and that of the second parallel to the axis of Z, Art. (76).

If $m'' = 0$ also, in the last two equations, the first will give

$$z = \pm\sqrt{-\frac{l}{p}}, \qquad x \textit{ and } y \textit{ indeterminate},$$

which represents two planes parallel to the plane XY, Art. (62); which are real when l and p have contrary signs; become one when $l = 0$; and are imaginary when l and p have the same sign; and *are particular cases of the cylinder*, analogous to the particular cases of the parabola discussed in Art. (171).

In the same way it may be proved, that the second equation will represent two planes parallel to the plane XZ.

If $m = 0$, $n = 0$, $p = 0$, the equation ceases to be one of the second degree.

From this discussion, we see that all surfaces of the second order will belong to one of the three classes represented by the following equations.

First, $$mx^2 + ny^2 + pz^2 + l = 0.$$

Second, $$ny^2 + pz^2 + m''x = 0.$$

Third, $$\left.\begin{array}{l} ny^2 + pz^2 + l = 0 \\ pz^2 + m''x + l = 0 \\ ny^2 + m''x + l = 0 \end{array}\right\}.$$

204. The centre of a surface is a point, through which if any straight line be drawn terminating in the surface, it will be bisected at this point.

If the origin of co-ordinates be placed at the centre, it is evident that for every point on one side of this origin, there must also be another in a directly opposite direction, at the same distance, and having the same co-ordinates with a contrary sign. Hence, the equation of the surface must be of such a form, that it will not change, when for $+x$, $+y$ and $+z$, $-x$, $-y$ and $-z$ are substituted; that is, all of its terms must be of an even degree with respect to the variables.

In order then to ascertain whether a given surface has a centre; we see if all the terms of its equation are of an even degree, if so, the origin of co-ordinates is a centre; if they are not, we then see if the origin of co-ordinates can be so placed as to make all the terms of the transformed equation, of an even degree. If this is possible, the surface will have a centre, which will be at the new origin. If it is not possible, the surface will have no centre.

205. By applying the above principles to surfaces of the second

order, we see that all of *the first class* have centres. That none of *the second* have centres. That the cylinders represented by the first equation of the third class have an infinite number of centres, each point of the axis fulfilling the required condition. That those represented by the second and third equations have no centres.

206. Any plane which bisects a system of parallel chords of a surface, is called *a diametral plane;* and if the chords are perpendicular to the plane, it is a principal diametral plane, or simply *a principal plane.*

Two diametral planes intersect in a diameter common to the two curves cut from the surface by these planes, and this intersection is also a diameter of the surface; and two principal planes intersect in an axis of the surface.

A diametral plane may be constructed, by drawing three parallel chords of the surface, not in the same plane, and bisecting them by a plane. By constructing two planes in this way, we determine a diameter, and the middle point of this diameter will evidently be the centre.

207. The co-ordinate planes being at right angles to each other, we see that each of them, in surfaces of the second order of the first class, is *a principal plane.* For, if equation (2), Art. (203), be solved with reference to either variable, we shall have two equal values with contrary signs, and these two values taken together, will form a chord, perpendicular to the co-ordinate plane of the other two variables, and bisected by it.

From this, it also follows that the axes of co-ordinates are axes of the surface, Art. (206).

In the second class, equation (4), Art. (203), the co-ordinate planes ZX and YX, are also principal planes, and the axis of X is an axis of the surface.

In the cylinders represented by the first equation of the third class, the planes ZX and YX are principal planes, and the axis of X is the axis of the cylinder.

In the cylinders represented by the second, the plane XY is the only principal plane, and there is no axis.

In those represented by the third, the plane ZX is the only principal plane, and there is no axis.

DISCUSSION OF THE VARIETIES OF SURFACES OF THE SECOND ORDER.

208. All the varieties of the first class of surfaces of the second order, or those which have a single centre, may be obtained by making in their equation, Art. (203).

First. m, n and p all positive, l being negative or positive.

Second. Either two positive and the other negative, l being positive.

Third. One positive and the other two negative, l being positive.

For if all are negative, the signs of both members of the equation may be changed, giving the first case.

If two are negative, the other positive and l negative, the signs may be changed, giving the second case.

If one is negative, the others positive, and l negative, the signs may be changed, giving the third case.

First, m, n and p positive, l negative or positive.

209. Supposing l to be negative, the equation of the first class, Art. (203), may be put under the form

$$mx^2 + ny^2 + pz^2 = l \text{..........} (1).$$

Let us intersect this surface by planes parallel, respectively, to the co-ordinate planes ZY, ZX and XY. The equations of the cutting planes, Art. (62), will be

$$x = h, \qquad y = k, \qquad z = g.$$

Combining these with the equation of the surface, Art. (62), we obtain

$$ny^2 + pz^2 = l - mh^2;$$
$$mx^2 + pz^2 = l - nk^2; \ldots\ldots\ldots\ldots(2),$$
$$mx^2 + ny^2 = l - pg^2;$$

for the equations of the projections of the several intersections on the co-ordinate planes; and since the curves are parallel to the planes on which they are projected, the projections are equal to the curves themselves.

Each of these equations represents an ellipse, Art. (169), and these ellipses will be real when the second members of the equations are positive, or

$$h < \pm\sqrt{\frac{l}{m}}, \qquad k < \pm\sqrt{\frac{l}{n}}, \qquad g < \pm\sqrt{\frac{l}{p}}.$$

If

$$h = \pm\sqrt{\frac{l}{m}}, \qquad k = \pm\sqrt{\frac{l}{n}}, \qquad g = \pm\sqrt{\frac{l}{p}},$$

the above equations reduce to

$$ny^2 + pz^2 = 0, \quad mx^2 + pz^2 = 0, \quad mx^2 + ny^2 = 0,$$

and the first members of each being the sum of two positive quantities, they can only be satisfied by making

$$y = 0, \ z = 0; \quad x = 0, \ z = 0; \quad x = 0, \ y = 0,$$

which are the equations of points

If

$$h > \pm\sqrt{\frac{l}{m}}, \qquad k > \pm\sqrt{\frac{l}{n}}, \qquad g > \pm\sqrt{\frac{l}{p}},$$

the second members of the above equations (2) will be negative, and they can be satisfied by no values of the variables, and the ellipses will be imaginary, that is, the planes will not intersect the surface.

If

$$h = 0, \qquad k = 0, \qquad g = 0,$$

equations (2), become

$$ny^2 + pz^2 = l, \qquad mx^2 + pz^2 = l, \qquad mx^2 + ny^2 = l,$$

which are the equations of the principal sections, and each of these sections is evidently larger than any other made by a parallel plane.

From this discussion we conclude that if the surface, represented by equation (1), be intersected by a system of planes parallel, respectively, to the co-ordinate planes, *the curves of intersection will all be ellipses*, and these ellipses will diminish as the distance of the cutting plane from the centre, on either side, is increased, until they reduce to points; after which there will be no intersection and no points of the surface. The surface is then limited in all directions, as in the figure, and is called *an Ellipsoid.*

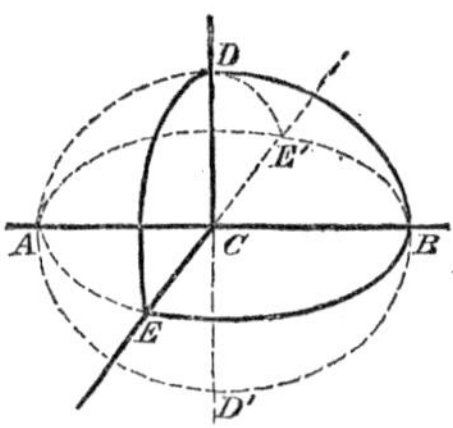

If we make $y = 0$, $z = 0$, in equation (1), we have

$$mx^2 = l, \qquad \text{or} \qquad x = \pm\sqrt{\frac{l}{m}} = \text{CB or CA.}$$

In a similar way we find

$$\pm\sqrt{\frac{l}{n}} = \text{CE or CE}', \qquad \pm\sqrt{\frac{l}{p}} = \text{CD or CD}'.$$

Placing the expressions for these semi-axes, respectively equal to a, b and c, we have

$$\sqrt{\frac{l}{m}} = a, \quad \sqrt{\frac{l}{n}} = b, \quad \sqrt{\frac{l}{p}} = c,$$

whence

$$m = \frac{l}{a^2}, \qquad n = \frac{l}{b^2}, \qquad p = \frac{l}{c^2}$$

and substituting these in equation (1), we obtain

$$\frac{x^2}{a^2} + \frac{y^2}{b^2} + \frac{z^2}{c^2} = 1,$$

or

$$b^2c^2x^2 + a^2c^2y^2 + a^2b^2z^2 = a^2b^2c^2,$$

an equation for the ellipsoid, referred to its centre and axes, analogous to equation (e), Art. (105).

210. If $m = n$, equation (1) of the preceding article may be put under the form

$$x^2 + y^2 = \frac{l - pz^2}{m} = \overline{f(z)}^2,$$

which is the equation of a surface of revolution, the axis of Z being the axis of revolution, Art. (198). But since $m = n$, we have $a = b$ or $\text{CA} = \text{CE}$, and the surface is generated by revolving the ellipse BDA about its conjugate axis, and is *the oblate spheroid*, Art. (198).

Likewise if $n = p$, equation (1), becomes

$$y^2 + z^2 = \frac{l - mx^2}{n} = \overline{f(x)}^2,$$

which is the equation of *the prolate spheroid.*

If $m = n = p$, we obtain

$$x^2 + y^2 + z^2 = \frac{l}{m},$$

which is *the equation of the sphere*, Art. (198).

If $l = 0$, equation (1) becomes

$$mx^2 + ny^2 + pz^2 = 0,$$

which, since the first member is the sum of three positive quantities, can only be satisfied by making

$$x = 0, \qquad y = 0, \qquad z = 0,$$

which are *the equations of a point*, Art. (41).

If l is positive, equation (2), Art. (203), takes the form

$$mx^2 + ny^2 + pz^2 = -l,$$

which can be satisfied for no values of x, y and z, and therfore represents no surface, or *an imaginary surface.*

From this discussion we see that the particular cases of the Ellipsoid are, the ellipsoid of revolution, the sphere, the point, and the imaginary surface.

Second, m and n positive, p negative and l positive.

211. In this case equation (2), Art. (203), takes the form

$$mx^2 + ny^2 - pz^2 = -l \ldots\ldots\ldots\ldots(1).$$

Intersecting the surface by planes as in Art. (209), we have, for the equations of the projections of the curves of intersection,

$$ny^2 - pz^2 = -l - mh^2;$$
$$mx^2 - pz^2 = -l - nk^2; \ldots\ldots\ldots\ldots(2)$$
$$mx^2 + ny^2 = -l + pg^2.$$

Each of the first two of these equations represents an hyperbola, whose transverse axis coincides with the axis of Z, Art. (105), and which increases in length, indefinitely as h and k increase.

The third equation represents an ellipse, Art. (105), which is real when

$$pg^2 > l, \qquad \text{or} \qquad g > \pm\sqrt{\frac{l}{p}},$$

and which increases as g increases. This ellipse becomes a point when

$$pg^2 = l, \qquad \text{or} \qquad g = \pm\sqrt{\frac{l}{p}} \text{.............}(3),$$

and imaginary, or there is no curve, when

$$pg^2 < l, \qquad \text{or} \qquad g < \pm\sqrt{\frac{l}{p}}.$$

If $h = 0$, $k = 0$, $g = 0$, Art. (62), equations (2), become

$$ny^2 - pz^2 = -l, \quad mx^2 - pz^2 = -l, \quad mx^2 + ny^2 = -l,$$

which are the equations of the principal sections.

The first two represent hyperbolas, whose transverse axes are less than those of any of the parallel hyperbolas. The third equation can be satisfied by no values of x and y, from which it appears that the plane XY does not intersect the surface.

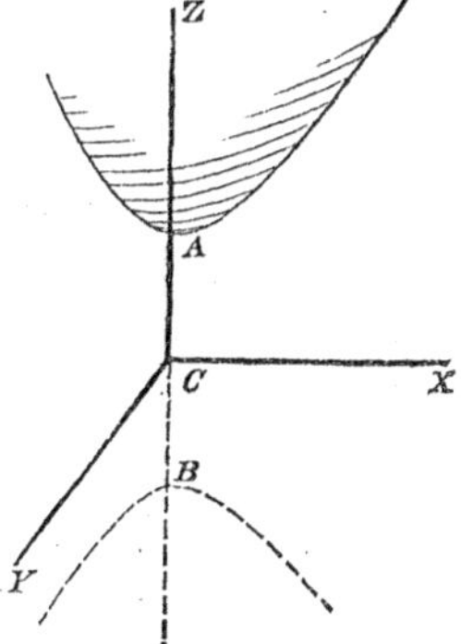

From this discussion, we conclude, that if the surface represented by equation (1) be intersected by a system of planes, parallel respectively to the co-ordinate planes, the sections parallel to ZX and ZY, will be hyperbolas having their transverse axes parallel to the axis of Z, while the sections parallel to XY, will be ellipses when at a greater distance from the origin, above or below, than the value of g in equation (3). Hence, the surface extends to infinity in all directions from the centre, and consists of two distinct and equal parts or nappes, as in the figure. It is therefore called, *an Hyperboloid of two nappes.*

If we make $y = 0, \quad z = 0,$ in equation (1), we have

$$mx^2 = -l, \qquad x = \pm\sqrt{\frac{-l}{m}},$$

which is imaginary, and the surface does not cut the axis of X.

In a similar way, we find

$$ny^2 = -l, \qquad y = \pm\sqrt{\frac{-l}{n}},$$

and

$$pz^2 = l, \qquad z = \pm\sqrt{\frac{l}{p}} = \text{CA or CB.}$$

Placing

$$\sqrt{\frac{-l}{m}} = c\sqrt{-1}, \quad \sqrt{\frac{-l}{n}} = b\sqrt{-1}, \quad \sqrt{\frac{l}{p}} = a,$$

we have

$$m = \frac{l}{c^2}, \qquad n = \frac{l}{b^2}, \qquad p = \frac{l}{a^2},$$

and these in equation (1), give

$$\frac{x^2}{c^2} + \frac{y^2}{b^2} - \frac{z^2}{a^2} = -1,$$

or

$$a^2b^2x^2 + a^2c^2y^2 - b^2c^2z^2 = -a^2b^2c^2.$$

for the equation of the hyperboloid of two nappes, referred to its centre and axes.

212. If $m = n$, equation (1) of the preceding article may be put under the form

$$x^2 + y^2 = -\frac{l - pz^2}{m} = \overline{f(z)}^2,$$

which is the equation of a surface of revolution, Art. (198), evidently generated by revolving the hyperbola about its transverse axis BA, or *the hyperboloid of revolution of two nappes.*

If $l = 0$, equation (1) reduces to

$$mx^2 + ny^2 - pz^2 = 0.$$

If this surface be intersected by any plane parallel to XY, we have for the projection of the intersection

$$mx^2 + ny^2 = pg^2,$$

which is the equation of an ellipse always real, whether g be positive or negative. If $g = 0$, we have

$$mx^2 + ny^2 = 0,$$

which can only be satisfied by

$$y = 0, \qquad x = 0,$$

which are the equations of a point. If we make first $x = 0$, and then $y = 0$, we obtain for the intersections by the co-ordinate planes YZ and XZ, the equations

$$ny^2 - pz^2 = 0, \qquad mx^2 - pz^2 = 0,$$

or

$$y = \pm z\sqrt{\frac{p}{n}}, \qquad x = \pm z\sqrt{\frac{p}{m}}$$

each of which evidently represents two right lines passing through the origin, Art. (169), and the surface can only be *a cone having its vertex at the origin*, and axis coinciding with the axis of Z.

The particular cases of the Hyperboloid of two nappes are, therefore, *the hyperboloid of revolution of two nappes, and the cone.*

Third, m positive, n and p negative, l positive.

213. In this case, equation (2), Art (203), takes the form

$$mx^2 - ny^2 - pz^2 = -l \text{............}(1).$$

Intersecting by planes, as in Art. (209), we obtain

$$ny^2 + pz^2 = l + mh^2.$$

$$mx^2 - pz^2 = -l + nk^2 \text{............}(2).$$

$$mx^2 - ny^2 = -l + pg^2.$$

The first of these equations represents an ellipse, which is always real, and increases as h increases in either direction, from the origin.

The second represents an hyperbola, whose transverse axis coincides with the axis of Z when the second member is negative, or

$$nk^2 < l, \qquad \text{and} \qquad k < \pm\sqrt{\frac{l}{n}},$$

and with the axis of X, when

$$k > \pm\sqrt{\frac{l}{n}},$$

The third is also the equation of an hyperbola, whose transverse axis coincides with the axis of Y, when

$$g < \pm\sqrt{\frac{l}{p}},$$

and with the axis of X, when

$$g > \pm\sqrt{\frac{l}{p}}.$$

If in the last two of equations (2), we make

$$k = \pm\sqrt{\frac{l}{n}}, \qquad g = \pm\sqrt{\frac{l}{p}},$$

we have

$$mx^2 - pz^2 = 0, \qquad x = \pm z\sqrt{\frac{p}{m}},$$

$$mx^2 - ny^2 = 0, \qquad x = \pm y\sqrt{\frac{n}{m}};$$

each of which represents two right lines.

If $h = 0$, $k = 0$, $g = 0$, equations (2) become

$$ny^2 + pz^2 = l, \qquad mx^2 - pz^2 = -l, \qquad mx^2 - ny^2 = -l,$$

for the equations of the principal sections.

The first represents an ellipse, which is smaller than any parallel section, and is called *the ellipse of the gorge.* The other two represent hyperbolas. We therefore conclude that, if the surface be intersected by planes parallel respectively to the co-ordinate planes, the sections parallel to ZX and YX are hyperbolas; while those parallel to YZ are ellipses, always real, whatever be their distances on either side of the centre. The surface then extends to infinity in all directions from the centre, without being separated into two parts. It is called *an hyperboloid of one nappe.*

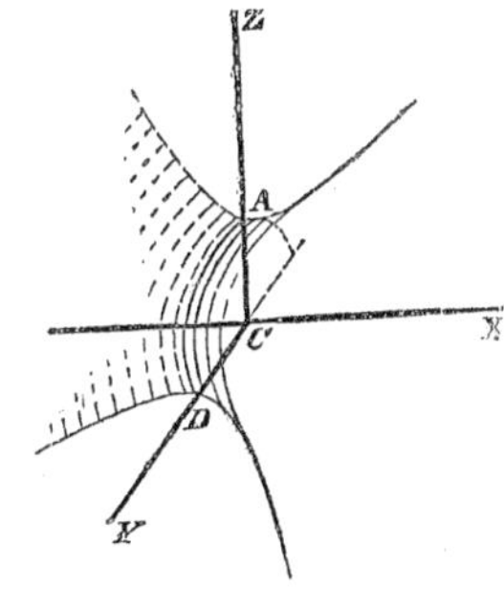

If we make $y = 0$, $z = 0$, in equation (1), we have

$$mx^2 = -l, \qquad x = \pm\sqrt{\frac{-l}{m}},$$

which is imaginary. In a similar way, we find

$$\text{CD} = \sqrt{\frac{l}{n}}, \qquad \text{CA} = \sqrt{\frac{l}{p}},$$

both of which are real. Placing

$$\sqrt{\frac{-l}{m}} = c\sqrt{-1}, \qquad \sqrt{\frac{l}{n}} = b, \qquad \sqrt{\frac{l}{p}} = a,$$

we deduce

$$m = \frac{l}{c^2}, \qquad n = \frac{l}{b^2}, \qquad p = \frac{l}{a^2},$$

and these in equation (1), give

$$a^2b^2x^2 - a^2c^2y^2 - b^2c^2z^2 = -a^2b^2c^2,$$

for the equation of the hyperboloid of one nappe, referred to its centre and axes.

214. If $n = p$, equation (1) of the preceding article may be written

$$y^2 + z^2 = \frac{l + mx^2}{n} = \overline{f(x)}^2,$$

which is the equation of a surface of revolution, Art. (198), evidently generated by revolving the hyperbola about its conjugate axis, or *the hyperboloid of revolution of one nappe.*

If $l = 0$, equation (1) reduces to

$$mx^2 - ny^2 - pz^2 = 0,$$

which may be shown, as in Art. (212), to be *the equation of a cone* having its vertex at the origin, and its axis coinciding with the axis of X.

The particular cases of the Hyperboloid of one nappe are, therefore, *the hyperboloid of revolution of one nappe, and the cone.*

215. All the varieties of the second class of surfaces of the second order, or those which have no centre, may be obtained by making in equation (4), Art. (203):

First, n and p positive, m'' being positive or negative:

Second, n positive and p negative, m'' being positive or negative.

For, if n and p are negative, the signs of both members of the equation may be changed giving the first case.

If n is negative and p positive, the signs may be changed giving the second case.

First, n and p positive, m'' positive or negative.

216. If m'' is negative, equation (4), Art. (203), may be put under the form

$$ny^2 + pz^2 = m''x \text{............}(1).$$

Intersecting the surface as in Art. (209), we have for the projections of the several curves on the co-ordinate planes,

$$ny^2 + pz^2 = m''h, \qquad pz^2 = m''x - nk^2, \qquad ny^2 = m''x - pg^2.$$

The first represents an ellipse, which is real as long as h is positive, and increases indefinitely as h is increased, becomes a point when $h = 0$, and is imaginary for all negative values of h. The other two represent parabolas, the axes of which coincide with the axis of X, Art. (84). And since the parameters of these parabolas are, respectively, $\frac{m''}{p}$ and $\frac{m''}{n}$, whatever be the values of k and g, it follows that all the parallel sections are equal to each other.

By making $h = 0$, $k = 0$, $g = 0$, we have for the principal sections

$$ny^2 + pz^2 = 0, \qquad pz^2 = m''x, \qquad ny^2 = m''x.$$

The first represents a point, the origin of co-ordinates, and each of the others a parabola, having its vertex at the origin.

From this it appears that the surface extends to infinity in the positive direction of the axis of X, but does not extend at all to the left of the origin; that the intersections by one system of planes are ellipses, and by the other two, parabolas. It is therefore called *an elliptical paraboloid*.

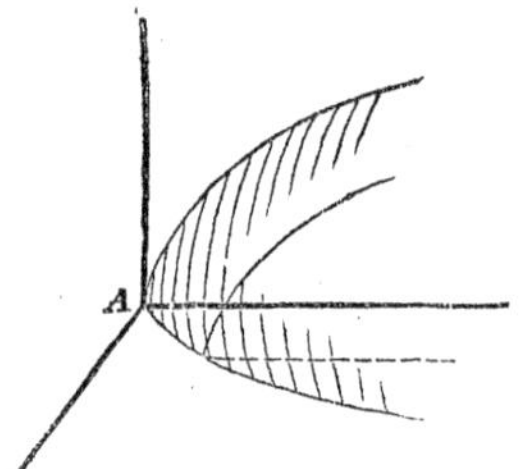

If m'' is positive, equation (4), Art. (203), takes the form

$$ny^2 + pz^2 = -m''x,$$

in which, if we change x into $-x$, we shall have equation (1). But the only effect of this change is to estimate the abscissas from A to the left. The equation will then represent the same surface revolved 180° about the axis of Y.

217. If $n = p$, equation (1) of the preceding article may be written

$$y^2 + z^2 = \frac{m''}{n}x = \overline{fx}^2,$$

which is the equation of *a paraboloid of revolution*, generated by revolving the parabola about its axis, and this is the only particular case of the elliptical paraboloid.

Second, n positive and p negative, m'' positive or negative.

218. It will only be necessary to discuss the case where m'' is negative; for, if m'' is positive, it may be shown, as in Art. (216), that the equation will represent the same surface revolved 180° about the axis of Y.

This being the case, equation (4), Art. (203), takes the form

$$ny^2 - pz^2 = m''x \ldots\ldots\ldots\ldots (1).$$

Intersecting the surface, as in Art. (216), we have

$$ny^2 - pz^2 = m''h \ldots (2), \quad pz^2 = -m''x + nk^2, \quad ny^2 = m''x + pg^2.$$

The first is the equation of an hyperbola always real, and having its transverse axis on the axis of Y when h is positive, and on the axis of Z when h is negative, Art. (105). The other two are the equations of parabolas, the first extending indefinitely in the direction of the negative abscissas, and the second in the direction of the positive abscissas, Art. (171).

By making $h = 0, \quad k = 0, \quad g = 0,$ we have for the principal sections

$$ny^2 - pz^2 = 0 \ldots\ldots (3), \qquad pz^2 = -m''x, \qquad ny^2 = m''x.$$

The first may be put under the form

$$ny^2 = pz^2, \qquad \text{or} \qquad y = \pm z\sqrt{\frac{p}{n}},$$

which represents two right lines passing through the origin. The other two represent parabolas each equal to those cut out by the corresponding parallel planes.

From this, it appears that the surface is unlimited in all directions; that the sections by one system of planes are hyperbolas, and by the other two, parabolas. It is therefore called *a hyperbolic paraboloid.*

It has no particular case.

We have seen above that the plane YZ intersects the surface in two right lines represented by equation (3), and that any plane parallel to YZ, intersects the surface in an hyperbola, the projection of which is represented by equation (2). If we denote the ordinate of any point of one of these right lines by y', to distinguish it from the ordinate of a point of the curve corresponding to the same value of z, we shall have

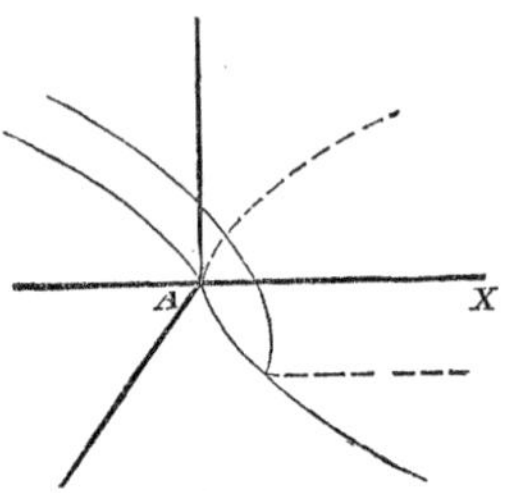

$$ny'^2 - pz^2 = 0.$$

Subtracting this equation, member by member, from equation (2), we have

$$ny^2 - ny'^2 = m''h; \qquad \text{whence} \qquad y - y' = \frac{m''h}{n(y + y')}.$$

Now as z is increased, y and y' are both increased, and $y - y'$ becomes smaller and smaller, and when y and y' become infinite, $y - y'$ becomes 0, or the two points coincide; that is, the right line continually approaches the curve and touches it at an infinite distance, or is *an asymptote*, Art. (161). Hence, the two right lines represented by equation (3), will be the asymptotes of the projections of the hyperbolas cut out by the planes parallel to YZ. Or, if two planes be passed through these lines and the axis of X, the plane which cuts from the surface an hyperbola, will cut from these planes, lines which will be the asymptotes of the hyperbola.

OF THE INTERSECTION OF SURFACES OF THE SECOND ORDER BY PLANES.

219. It has been proved, Art. (200), that every intersection of a surface of the second order, by a plane, is a line of the second

order. The discussion of the nature of these sections, except when they are parallel to one of the co-ordinate planes, is much simplified by referring them to axes at right angles, in their own planes.

For the purpose of this discussion, let us resume the general equation, Art. (202),

$$mx^2 + ny^2 + pz^2 + m''x + l = 0............(1),$$

in which the origin is at some point A, on the line AX, this being the intersection of two principal planes, Art. (206). Let any plane be passed intersecting the surface, and let A′X′ be its trace on the plane XY, making an angle β with the axis of X, and let θ denote the angle made by this plane with the plane XY.

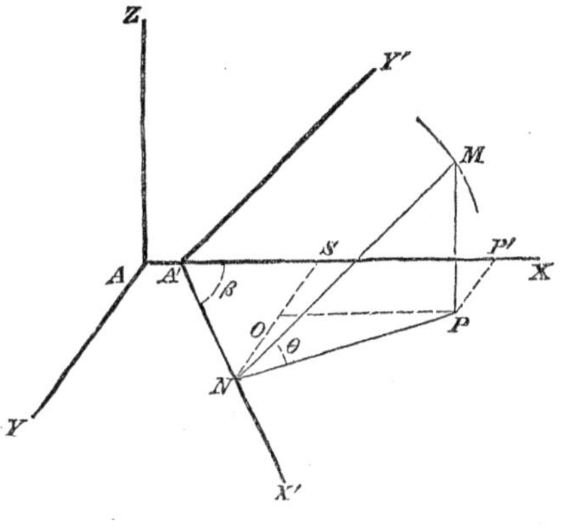

For any point of the curve of intersection, as M, we shall then have

$$x = \text{AP}', \qquad y = \text{PP}', \qquad z = \text{MP}.$$

Let this point be now referred to the two axes A′X′ and A′Y′, at right angles to each other and in the plane of the curve. Through P draw PN perpendicular to A′X′, and PO parallel to AX; also draw NS perpendicular to AX. Join M and N, then the angle MNP $= \theta$. Denote the distances

AA′ by a', A′N by x', MN by y',

and we shall have

$$x = a' + \text{A'S} + \text{OP}, \qquad y = \text{NS} - \text{NO}.$$

The right angled triangles MPN, A′SN, and PON, give

$$z = y' \sin\theta, \qquad \text{NP} = y' \cos\theta, \qquad \text{A'S} = x' \cos\beta,$$

$$\text{NS} = x' \sin\beta, \qquad \text{NO} = \text{NP} \cos\beta, \qquad \text{PO} = \text{NP} \sin\beta.$$

Substituting these values in the preceding equations, we obtain

$$x = a' + x' \cos\beta + y' \cos\theta \sin\beta, \qquad y = x' \sin\beta - y' \cos\theta \cos\beta.$$

If these values, with the value, $z = y' \sin\theta$, be substituted in equation (1), the result can only belong to points common to the plane and surface, and will therefore represent the line of intersection. Making the substitution and reducing, we obtain

$$(m \cos^2\beta + n \sin^2\beta)x'^2 + [\cos^2\theta(m \sin^2\beta + n \cos^2\beta) + p \sin^2\theta]\, y'^2$$
$$+ 2(m - n) \sin\beta \cos\beta \cos\theta\, x'y' + \cos\beta(2ma' + m'')x'$$
$$+ \cos\theta \sin\beta(2a'm + m'')y' + ma'^2 + m''a' + l = 0 \ldots (2).$$

By assigning proper values to a', we may always cause the plane to intersect the surface, and by assigning proper values to β and θ, we may cause the above equation to represent the several varieties of lines of the second order.

220. For instance, let it be required that the intersection shall be a right line or lines.

If it is possible to cut a right line from the surface by a plane in any position, the same right line may be cut out by a plane perpendicular to the plane XY. For it is only necessary that the cutting plane should occupy the position of the plane which projects the line on the co-ordinate plane XY. We may therefore regard θ in the above equation as equal to 90°, which gives

$$\cos\theta = 0, \qquad \sin\theta = 1,$$

and see if it is possible to give such values to a' and β, as will make the equation represent one or more right lines.

221 For those surfaces which have a centre, we may also regard $m'' = 0$, Art. (203). Substituting this value with the above, for $\cos\theta$ and $\sin\theta$, in equation (2), Art. (219), and omitting the dashes of x and y, it reduces to

$$(m\cos^2\beta + n\sin^2\beta)x^2 + py^2 + 2a'm\cos\beta x + ma'^2 + l = 0.$$

Solving this with reference to y, we have

$$y = \pm\sqrt{-\frac{1}{p}[(m\cos^2\beta + n\sin^2\beta)x^2 + 2ma'\cos\beta x + ma'^2 + l]} \ldots (1).$$

In order that this represent one or more real right lines, it is necessary that $-\frac{1}{p}$ shall be positive, and that the factor within the parenthesis shall be a perfect square, Art. (178), which requires

$$p < 0 \ldots\ldots\ldots\ldots (2),$$

and

$$(m\cos^2\beta + n\sin^2\beta)(ma'^2 + l) = m^2a'^2\cos^2\beta \ldots\ldots\ldots\ldots (3).$$

Deducing the value of a' from the last condition, we obtain

$$a' = \pm\sqrt{-\frac{l(m\cos^2\beta + n\sin^2\beta)}{mn\sin^2\beta}} \ldots\ldots\ldots\ldots (4).$$

Since p is positive in the ellipsoid, Art. (209), condition (2) can not be fulfilled; whence the conclusion, that *no right line can be cut from this surface.*

Since m, n and l are positive in the hyperboloid of two nappes, Art. (211), the value of a' will be imaginary for all values of β. Condition (3) can not then be fulfilled, and *no right line can be cut from this surface.*

Since m and l are positive and n and p negative in the hyperboloid of one nappe, Art. (213), condition (2) will be fulfilled, and the values of a' will be real for all values of β which give

$$n \sin^2 \beta < m \cos^2 \beta, \qquad \text{or} \qquad \tang^2 \beta < \frac{m}{n},$$

and equation (1) will then represent two real right lines which intersect, Art. (178). Hence, an infinite number of right lines may be cut from the surface of the hyperboloid of one nappe by planes.

222. If we take the value of $\sqrt{ma'^2 + l}$ from condition (3) of the preceding article, and substitute it in equation (1), extract the square root of the factor within the parenthesis, and substitute in the result the value of a', from equation (4), we shall obtain

$$y = \pm\left(\frac{x}{\sqrt{-p}}\sqrt{m \cos^2 \beta + n \sin^2 \beta} + \frac{\cos \beta}{\sin \beta}\sqrt{\frac{ml}{pn}}\right),$$

which may be put under the form

$$y = \pm\left(\frac{x \sin \beta}{\sqrt{-p}}\sqrt{m \cot^2\beta + n} + \cot \beta \sqrt{\frac{ml}{pn}}\right)......(1),$$

and will represent the two right lines cut out by any plane making the angle β with the plane XZ. By changing β into β', we shall obtain at once the equations of two other lines cut out by another plane. The lines cut out by two different planes are not parallel; for the cutting planes, which are also their projecting planes, are not parallel. Neither can they intersect, for if they

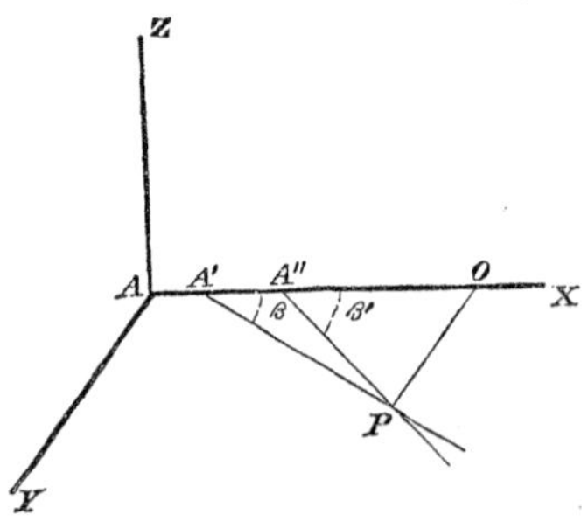

intersect at all, it must be in the perpendicular to the plane XY, at the point P; and if we substitute A'P for x in the equation of the first set of lines, and A″P for x in the equation of the second set, we must obtain the same value for y in each case. But denoting the distance PO by d, we have

$$A'P = \frac{d}{\sin\beta}, \qquad A''P = \frac{d}{\sin\beta'},$$

and these values being substituted for x, each in equation (1), will give values which are unequal.

223. For those surfaces which have no centre, we may regard m and l as equal to 0, Art. (203). Substituting these values with $\cos\theta = 0$, $\sin\theta = 1$, in equation (2), Art. (219), and omitting the dashes of the variables x and y, it reduces to

$$n\sin^2\beta x^2 + py^2 + m''\cos\beta x + m''a' = 0.$$

Solving this with reference to y, we have

$$y = \pm\sqrt{-\frac{1}{p}(n\sin^2\beta x^2 + m''\cos\beta x + m''a')} \ldots\ldots\ldots\ldots(1).$$

In order that this shall represent one or more real right lines, we must have, as in Art. (221),

$$p < 0 \ldots\ldots\ldots\ldots(2),$$

and

$$m''^2\cos^2\beta = 4n\sin^2\beta\, m''a' \ldots\ldots\ldots\ldots(3);$$

whence

$$a' = \frac{m''\cos^2\beta}{4n\sin^2\beta}.$$

In the elliptical paraboloid, n and p are both positive, Art. (216), condition (2) can not be fulfilled, and *no right line can be cut from the surface.*

In the hyperbolic paraboloid, n is positive and p negative, Art. (218); condition (2) is fulfilled, the value of a' will be real, and fulfil condition (3) for all values of β; and an infinite number of right lines may be cut from this surface by planes. Substituting the value of a' in equation (1), and extracting the square root of the quantity within the parenthesis, we obtain

$$y = \pm \frac{\sqrt{n}}{\sqrt{-p}}\left(\sin \beta x + \frac{m'' \cos \beta}{2n \sin \beta}\right),$$

which will represent the two right lines cut out by any plane making the angle β with the plane XZ. By changing β into β', we shall obtain the equations of two other right lines cut out by another plane.

It may be proved, as in the preceding article, that the lines cut out by two different planes are not parallel, and do not intersect.

224. The preceding discussion of the rectilineal sections of surfaces of the second order, enables us to classify these surfaces as they are classed in Descriptive Geometry. This classification is:

1. *Plane surfaces*, which may be generated by a right line moving along another right line and parallel to its first position.

2. *Single curved surfaces*, which may be generated by a right line, moving so that its consecutive positions shall be in the same plane.

3. *Double curved surfaces*, which can only be generated by curves.

4. *Warped surfaces*, which may be generated by a right line, moving so that its consecutive positions shall not be in the same plane.

The cylindrical and conical surfaces are *single curved*, as the consecutive elements of the first are parallel, Art. (74), and those of the second intersect, Art. (77); that is, are in the same plane.

The ellipsoid, hyperboloid of two nappes, and elliptical paraboloid, are *double curved;* since no right line can be cut from them, Arts. (221), (223); that is, no right line can be so placed as to lie wholly in either surface.

The hyperboloid of one nappe, and hyperbolic paraboloid, are *warped;* since the right lines cut from the surfaces by consecutive planes are not parallel, neither do they intersect, Arts. (222), (223), and therefore can not lie in the same plane.

225. If it be required that the intersection represented by equation (2), Art. (219), shall be a circle, it is necessary that the coefficient of $x'y'$ be equal to 0, and that the coefficients of x'^2 and y'^2 be equal to each other, Art. (169). This requires

$$2(m - n) \sin \beta \cos \beta \cos \theta = 0 \ldots\ldots\ldots\ldots (1),$$

$$m \cos^2 \beta + n \sin^2 \beta = \cos^2 \theta (m \sin^2 \beta + n \cos^2 \beta) + p \sin^2 \theta \ldots\ldots (2).$$

The condition (1), (m and n being in general unequal), may be satisfied by making either

$$\sin \beta = 0, \qquad \cos \beta = 0, \qquad \cos \theta = 0.$$

Sin $\beta = 0$ substituted in condition (2), gives

$$m = n \cos^2 \theta + p \sin^2 \theta = m(\sin^2 \theta + \cos^2 \theta),$$

since $\sin^2 \theta + \cos^2 \theta = 1$. From this we deduce

$$m \operatorname{tang}^2 \theta + m = n + p \operatorname{tang}^2 \theta,$$

or

$$\operatorname{tang} \theta = \pm \sqrt{\frac{n - m}{m - p}} \ldots\ldots\ldots\ldots (3).$$

In a similar way we find

$$\cos \beta = 0, \qquad \operatorname{tang} \theta = \pm \sqrt{\frac{n - m}{p - n}} \ldots\ldots\ldots\ldots (4);$$

$$\cos \theta = 0, \qquad \text{tang}\, \beta = \pm\sqrt{\frac{p - m}{n - p}} \ldots\ldots\ldots\ldots (5).$$

In the first case, the cutting plane is parallel to the axis of X; in the second, parallel to the axis of Y; and in the third, parallel to the axis of Z.

It may be remarked that if any position of the cutting plane be found to give a circle, every parallel plane intersecting the surface will also give a circle. For if the angles β and θ remain the same, a' may be changed at pleasure, without affecting the equality of the coefficients of x'^2 and y'^2.

226. *In the ellipsoid*, in which m, n and p are positive, Art. (209), in order that the first set of values of tang θ, (equation (3), preceding article), may be real, we must have

$$n > m > p, \qquad \text{or} \qquad p > m > n.$$

In order that the second set may be real, we must have

$$p > n > m, \qquad \text{or} \qquad m > n > p.$$

In order that the values of tang β may be real, we must have

$$n > p > m, \qquad \text{or} \qquad m > p > n.$$

It is evident that no two of these conditions can be fulfilled at the same time.

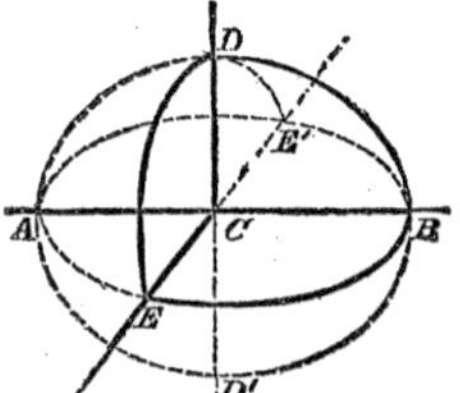

If either of the first is fulfilled, we shall have, [see expressions for a, b, and c, Art. (209)],

$$\text{CE} > \text{CB} > \text{CD}, \text{ or } \text{CE} < \text{CB} < \text{CD}.$$

Hence, the mean axis of the surface lies on the axis of X, to which the cutting plane is parallel.

If either of the second is fulfilled, we shall have

$$CB > CE > CD \qquad \text{or} \qquad CB < CE < CD,$$

and for either of the third

$$CB > CD > CE, \qquad \text{or} \qquad CB < CD < CE.$$

Hence, a cutting plane passed parallel to the mean axis of the surface may have two positions, such that the sections shall be circles, these positions being determined by the two proper values of $\operatorname{tang}\theta$ or $\operatorname{tang}\beta$; and in no other position can the section be a circle.

If $m = n$, both sets of values of $\operatorname{tang}\theta$ become 0, and $\operatorname{tang}\beta$ becomes imaginary. Hence the two cutting planes unite in one, parallel to XY, or perpendicular to the axis of Z; as should be the case, since the surface becomes an ellipsoid of revolution, its axis lying on the axis of Z, Art. (210).

If $n = p$, the first set of values of $\operatorname{tang}\theta$ become imaginary, while the second and those of $\operatorname{tang}\beta$ become infinite, and the cutting plane is perpendicular to the axis of X, Art. (210).

If $m = n = p$, the values of $\operatorname{tang}\theta$ and $\operatorname{tang}\beta$ become $\frac{0}{0}$, indeterminate, and every position of the cutting plane gives a circle, as it should, since the surface becomes a sphere.

227. *In the hyperboloid of two nappes*, in which m and n are positive and p negative, the values of Art. (225), after giving to the letters their proper signs, become

$$\operatorname{tang}\theta = \pm\sqrt{\frac{n-m}{m+p}}, \qquad \operatorname{tang}\theta = \pm\sqrt{\frac{m-n}{p+n}},$$

$$\operatorname{tang}\beta = \pm\sqrt{-\frac{p+m}{n+p}}.$$

The values of $\operatorname{tang}\beta$ are imaginary.

If $m < n$, the first set of values of $\text{tang}\ \theta$ is real, and the second imaginary. If $m > n$, the reverse is the case. But, if $m < n$, we have $c > b$; and if $m > n$, we have $c < b$, Art. (211). Hence, in this surface, the cutting plane must be parallel to the longest of the two axes which do not intersect the surface.

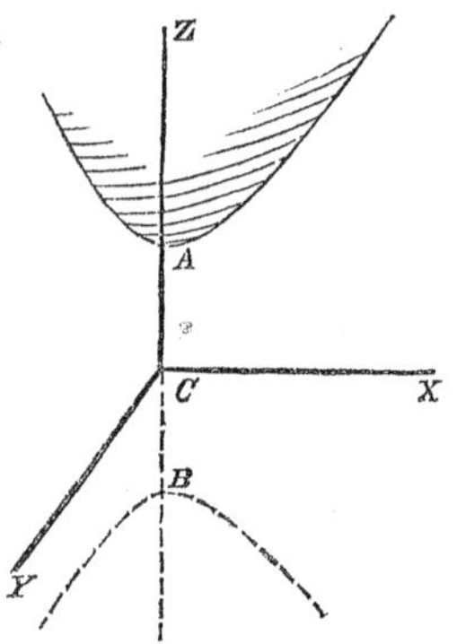

If $m = n$, the values of $\text{tang}\ \theta$ become 0, and the cutting planes unite in one perpendicular to the axis of Z; as they should, since in this case, we have the hyperboloid of revolution of two nappes, Art. (212).

Since the above values of $\text{tang}\ \theta$ do not depend upon l, they will remain the same when $l = 0$, Art. (212), that is, in a cone with an elliptical base, it is always possible to pass planes in two different directions so as to cut circles. These are called *sub-contrary sections.* If one of them be regarded as the base of the cone, the other will be sub-contrary to the base; that is, in a scalene cone with a circular base, *it is always possible to pass a system of planes not parallel to the base, which shall cut out circles.*

If the cone is a right cone with a circular base, it is a surface of revolution, and the sub-contrary sections unite in one, perpendicular to the axis or parallel to the base.

228. *In the hyperboloid of one nappe,* in which m is positive, n and p *negative,* we have

$$\text{tang}\ \theta = \pm\sqrt{-\frac{n+m}{m+p}}, \quad \text{tang}\ \theta = \pm\sqrt{-\frac{n+m}{n-p}},$$

$$\text{tang}\ \beta = \pm\sqrt{\frac{p+m}{n-p}}.$$

The first are imaginary.

If $n < p$, the second will be real and the third imaginary, and the reverse when $n > p$.

If $n < p$, we have

$$b = \text{CD} > a = \text{CA},$$

and if $n > p$, we have

$$\text{CD} < \text{CA}.$$

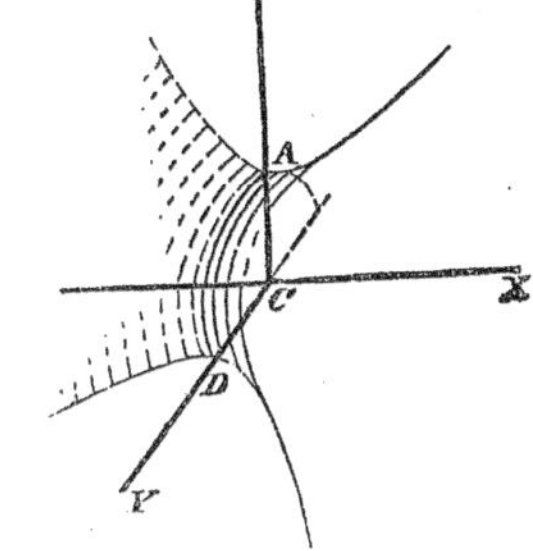

Hence, the cutting plane is parallel to the greatest of the two axes which pierce the surface.

If $n = p$, the above real values of tang θ and tang β become infinite and the two planes unite in one, perpendicular to the axis of X, Art. (214). When the surface becomes a cone, the discussion is similar to that in the preceding article.

229. *In the elliptical paraboloid*, in which $m = 0$, n and p positive, the values of tang θ and tang β become

$$\text{tang}\ \theta = \pm\sqrt{-\frac{n}{p}}, \qquad \text{tang}\ \theta = \pm\sqrt{\frac{n}{p-n}},$$

$$\text{tang}\ \beta = \pm\sqrt{\frac{p}{n-p}}.$$

The first are imaginary. If $n < p$, the second are real and the third imaginary. If $n > p$, the reverse will be the case. Hence, the cutting plane must be parallel to the greater axes of the elliptical sections, Art. (216).

If $n = p$, the above real values become infinite and the cutting planes unite in one perpendicular to the axis of X, Art. (217).

230. *In the hyperbolic paraboloid*, in which $m = 0$, n positive and p negative, we have

$$\text{tang } \theta = \pm\sqrt{\frac{n}{p}}, \qquad \text{tang } \theta = \pm\sqrt{-\frac{n}{p+n}},$$

$$\text{tang } \beta = \pm\sqrt{-\frac{p}{p+n}}.$$

The second and third are imaginary, and the first real, and the position of the cutting plane will be given by the equations

$$\sin \beta = 0, \qquad \text{tang } \theta = \pm\sqrt{\frac{n}{p}}$$

But these values with the value $m = 0$, substituted in condition (2), Art. (225), make the coefficients of x'^2 and y'^2 both equal to 0, and equation (2) of Art. (219), takes the form

$$ex + f = 0, \qquad y \textit{ indeterminate},$$

which represents a right line.

Since any plane parallel to either of the planes determined by the above values of $\sin \beta$ and $\text{tang } \theta$ will also cut a right line from the surface, we see that there are two different systems of right line elements, each of which is parallel to a given plane.

We conclude, also, *that no circle can be cut from the hyperbolic paraboloid.*

231. The intersection of any two surfaces of the second order may be found as in Art. (62); but as their equations are of the second degree, the result of their combination, so as to eliminate one of the variables, will be of the fourth degree; hence, in general, the projections of the lines of intersection will be lines of the fourth order, the discussion of which will be complicated and of little interest.

If, however, it is known that two such surfaces intersect in a line of the second order, it will, in general, be found that they will also intersect in another line of the second order; that is, *if one surface enters the other in a line of the second order, it will leave it in a line of the same order.*

To prove this, let us take the most general equations of the two surfaces,

$$mx^2 + ny^2 + pz^2 + m'xy + n'xz + p'yz$$
$$+ m''x + n''y + p''z + l = 0 \ldots\ldots(1),$$
$$qx^2 + ry^2 + sz^2 + q'xy + r'xz + s'yz$$
$$+ q''x + r''y + s''z + l' = 0 \ldots\ldots(2),$$

and let the plane of the curve in which it is known the two surfaces intersect be taken as the plane XY.

If we make $z = 0$, in each of the above equations, we shall have

$$mx^2 + ny^2 + m'xy + m''x + n''y + l = 0 \ldots\ldots\ldots\ldots(3);$$
$$qx^2 + ry^2 + q'xy + q''x + r''y + l' = 0 \ldots\ldots\ldots\ldots(4);$$

each of which must represent the known curve of intersection of the surfaces. These equations must then be the same, which can only be the case when the corresponding coefficients are equal, or when those of the first equation are equal to those of the second multiplied by a constant factor, as k. If we now multiply equation (2) by k and subtract from equation (1), we obtain

$$(p - ks)z^2 + (n' - kr')xz + (p' - ks')yz + (p'' - ks'')z = 0,$$

which equation must be satisfied for all values of x, y and z, belonging to points common to the two surfaces. Since z is a common factor, it may be satisfied by placing

$$z = 0,$$

or

$$(p - ks)z + (n' - kr')x + (p' - ks')y + (p'' - ks'') = 0,$$

$z = 0$ evidently belongs to the plane XY, in which the known line of intersection lies. The other is the equation of a plane, Art. (57), which by its combination with either (1) or (2), will give another line common to both surfaces, and this line must, of course, be one of the second order, Art. (200).

232. Let x'', y'' and z'' be the co-ordinates of a given point on a surface of the second order. These when substituted for the variables in the general equation

$$mx^2 + ny^2 + pz^2 + m''x + l = 0 \ldots\ldots\ldots\ldots (1),$$

must satisfy it, and give the equation of condition

$$mx''^2 + ny''^2 + pz''^2 + m''x'' + l = 0 \ldots\ldots\ldots\ldots (2).$$

Subtracting this, member by member, from equation (1), and factoring the terms, we have

$$m(x - x'')(x + x'') + n(y - y'')(y + y'')$$
$$+ p(z - z'')(z + z'') + m''(x - x'') = 0 \ldots\ldots\ldots\ldots (3),$$

which is the equation of the surface, with the condition introduced, that the point x'', y'', z'' shall be on the surface.

The equations of any right line passing through the given point, are

$$(x - x'') = a(z - z''), \qquad y - y'' = b(z - z'') \ldots\ldots\ldots\ldots (4).$$

If these equations be combined with equation (3), we shall obtain

$$(z - z'')[ma(x + x'') + nb(y + y'') + p(z + z'') + m''a] = 0 \ldots\ldots (5),$$

in which x, y and z must denote the co-ordinates of all points common to the line and surface, Art. (58). Since this is an equation of the second degree, there are but two such points; and these may be determined by placing the factors of (5) separately equal to 0.

$$z - z'' = 0, \quad \text{gives} \quad z = z'', \quad y = y'', \quad x = x'',$$

which evidently belong to the given point. Placing

$$ma(x + x'') + nb(y + y'') + p(z + z'') + m''a = 0 \ldots\ldots\ldots(6),$$

x, y and z in this must represent the co-ordinates of the second point in which the line pierces the surface.

If now any plane be passed through this right line, it will cut from the surface a line which will contain both of the points; and if the second point be moved along this line until it coincides with the first, the right line will become tangent to the line cut from the surface, and the values of x, y and z in equation (6), will become equal to x'', y'' and z''. Substituting these values in equation (6), it becomes

$$2max'' + 2nby'' + 2pz'' + m''a = 0 \ldots\ldots\ldots\ldots(7),$$

an equation which shows the relation that must exist between a and b, in order that the right line represented by equations (4) may be tangent to a line of the surface at the given point; and since a and b in this equation are indeterminate, it follows that an infinite number of right lines may be drawn, each tangent to a line of the surface at the given point.

If now in equation (7), we substitute for a and b their values taken from equations (4), we obtain

$$(2mx'' + m'')(x - x'') + 2ny''(y - y'') + 2pz''(z - z'') = 0,$$

or, since from equation (2),

$$-2mx''^2 - 2ny''^2 - 2pz''^2 = 2m''x'' + 2l,$$

we have finally,

$$2mx''x + 2ny''y + 2pz''z + m''(x + x'') + 2l = 0 \ldots\ldots\ldots\ldots(8),$$

an equation which expresses the relation between x, y and z for all points of the tangent line in all of its positions. The surface represented by it, is then the locus of all right lines drawn tangent to lines of the surface, at the given point, or point of contact. This equation being of the first degree between three variables, is the equation of a plane. This plane is said to be tangent to the surface, at the given point; and in general, *a plane is tangent to a surface, when it has at least one point in common with it, through which if any plane be passed, the sections made in the surface and plane will be tangent to each other.*

For those surfaces which have a centre, the origin of co-ordinates being placed at this centre, we have $m'' = 0$, Art. (203), and equation (8) reduces to

$$mxx'' + nyy'' + pzz'' + l = 0 \ldots\ldots\ldots\ldots (9).$$

If $m = n = p$, equation (9) becomes

$$xx'' + yy'' + zz'' = -\frac{l}{m} = R^2,$$

for the equation of a tangent plane to a sphere, Art. (210).

For those surfaces which have no centre, $m = 0$, $l = 0$, Art. (203), and equation (8) reduces to

$$2nyy'' + 2pzz'' + m''(x + x'') = 0.$$

233. Let x', y' and z' be the co-ordinates of a fixed point without a surface of the second order. If it be required that the tangent plane to the surface shall pass through this point, its co-ordinates must satisfy equation (8), of the preceding article, and give the equation of condition

$$2mx''x' + 2ny''y' + 2pz''z' + m''(x' + x'') + 2l = 0 \ldots\ldots\ldots (1).$$

In this equation x'', y'' and z'' are unknown, but since the point which they represent must lie on the surface, we must also have the condition

$$mx''^2 + ny''^2 + pz''^2 + m''x'' + l = 0 \ldots\ldots\ldots\ldots (2),$$

and these two equations are all the means which we have of determining the values of x'', y'' and z''; and since we thus have three unknown quantities, and but two equations, it follows that the unknown quantities are *indeterminate*. Hence we conclude that, in general, an infinite number of planes can be drawn from a point without a surface of the second order tangent to the surface.

If straight lines be drawn, from the different points of contact of these planes, to the fixed point, they will evidently form a cone which will be tangent to the surface, in the line formed by joining the points of contact. But since the co-ordinates of these points must all satisfy equation (1), when substituted for x'', y'' and z'', the points must lie in the plane which will be represented by this equation when x'', y'' and z'' are regarded as variables. This curve of contact must then be a plane curve, and since it lies on the surface at the same time, *it must be a line of the second order*, Art. (200). We therefore conclude that, in general, the line of contact of a tangent cone and surface of the second order, *is a line of the second order*. And the same will be true of a tangent cylinder, inasmuch, as the cone becomes a cylinder, when its vertex is removed to an infinite distance.

234. If it be required that the tangent plane pass through a second given point x''', y''', z''', without the surface, or contain the right line joining these two points, we shall also have the equation of condition

$$2mx''x''' + 2ny''y''' + 2pz''z''' + m''(x''' + x'') + 2l = 0,$$

and this united with equations (1) and (2) of the preceding article, will give three equations involving three unknown quantities, and since two of these equations are of the first, and the other of the second degree, there will in general be two sets of values for x'', y'' and z''. Hence we conclude that, in general, two planes may be passed through a right line tangent to a surface of the second order, and only two.

235. *A right line, or a plane, is normal to a surface when it is perpendicular to a tangent plane, at the point of contact.*

There evidently can be but one normal line to a surface at a given point; but, since every plane containing a normal will be perpendicular to the tangent plane, there will be an infinite number of normal planes.

236. The equations of a normal line, to a surface of the second order, will be of the form, Art. (50),

$$x - x'' = a(z - z''), \qquad y - y'' = b(z - z'') \ldots\ldots\ldots\ldots (1),$$

in which it is necessary to determine the values of a and b on condition that the line shall be perpendicular to the tangent plane represented by equation (8), Art. (232). The equations of condition, Art. (59),

$$a = -c, \qquad b = -d,$$

give

$$a = \frac{2mx'' + m''}{2pz''}, \qquad b = \frac{ny''}{pz''},$$

and these, in equations (1), give

$$x - x'' = \frac{2mx'' + m''}{2pz''}(z - z''), \qquad y - y'' = \frac{ny''}{pz''}(z - z'') \ldots (2),$$

for the equations of a normal line to any surface of the second order.

By supposing $m'' = 0$, we shall have the particular equations for those surfaces which have a centre; and by making $m = 0$, we have them for those surfaces which have no centre

If $n = p$, equation (2) reduces to

$$yz'' - y''z = 0,$$

which, having no absolute term, shows that the projection of the normal on the plane YZ passes through the origin of co-ordinates; hence the normal intersects the axis of X. But when $n = p$, the surface becomes one of revolution, the axis of X being the axis of revolution, Arts. (210), (214), (217). We therefore conclude that all the normals to a surface of revolution of the second order intersect the axis of revolution; and that the meridian plane, passing through the point of contact of a tangent plane, is a normal plane: Or, *a tangent plane to a surface of revolution of the second order is perpendicular to the meridian plane passing through the point of contact.*

PRACTICAL EXAMPLES.

237. Although examples have been occasionally given, in immediate connection with the articles which they are intended to illustrate, it is believed to be advantageous to add, in this place, a number of others, a portion of which the teacher may give out with each lesson; or may defer them until the subject has been completed, when their solution will serve as a general review of the principles of the course.

Each example should be carefully constructed, on the black board, in proper proportion, a unit of convenient length being first assumed; or, when it can be done, should be accurately drawn on paper, with mathematical instruments. By this exercise, the principles of the subject will be strongly impressed upon the mind of the pupil, while, at the same time, a good test of his knowledge will be afforded to his teacher.

The axes of co-ordinates are supposed to be at right angles, unless otherwise mentioned.

The teacher may multiply the examples to an unlimited extent, by simply substituting, for the numbers used, any others which may occur to him.

1. Construct the points whose equations are, Art. (16),

$$x = 2, \quad y = -1; \qquad x = -1, \quad y = 4;$$

$$x = -3, \quad y = -2; \qquad x = 3, \quad y = -5.$$

2. Find the expressions for the distances between the points, whose equations are, Art. (17),

$$x' = 1, \quad y' = 3; \qquad x'' = 0, \quad y'' = -2;$$

$$x' = -3, \quad y' = 4; \qquad x'' = 2, \quad y'' = -1.$$

3. Construct the points whose polar equations are, Art. (18),

$$v = 20^\circ, \quad r = 5; \qquad v = 190^\circ \quad r = 2.$$

4. Construct the right lines whose equations are, Art. (26),

$$2y - 3x + 1 = 0; \qquad 3y - x = 0.$$

5. Find the point of intersection of the right lines, whose equations are as in the last example, Art. (27).

6. Find the expression for the tangent of the angle, included by the same lines, Art. (28).

7. Ascertain if the lines represented by the equations

$$2y - 5x - 1 = 0, \qquad y = 3x - 2,$$

are parallel, or perpendicular to each other, Art. (28).

8. Find the equation of a right line, passing through the point $x' = 2$, $y' = -4$, and parallel to the line whose equation is, Art. (30),

$$3y + 2x - 1 = 0.$$

9. Find the equation of the right line passing through the same point and perpendicular to the same line, Art. (30); also, the length of the perpendicular, Art. (17).

10. Find the equation of a right line passing through the two points,

$$x' = 3, \quad y' = -4; \qquad x'' = -2, \quad y'' = -1.$$

11. Find and discuss the equation of a circle, the co-ordinates of whose centre are $x' = 3, \quad y' = -2$; and whose radius is 3, Art. (34).

Also, when the co-ordinates of the centre are $x = -2$, $y' = 0$; and the radius 4.

12. Find the intersection of the last circle with the right line whose equation is, Art. (27).

$$y = -3x - 1.$$

13. Ascertain if the point $x = 1, \quad y = -2$, is on, without, or within the circle whose equation is, Art. (37),

$$x^2 + y^2 = 9.$$

14. Find the equation of a circle which shall pass through the point $x = 3, \quad y = -2$; the origin being at the centre.

Also of one passing through the point $x = 4, \quad y = 2$, the origin being at the left hand extremity of the diameter, Art. (38).

15. Construct the points, in space, Art. (40);

$$x = 1, \qquad y = 2, \qquad z = 3;$$

$$x = -2, \qquad y = 3, \qquad z = -4.$$

16. Find the expression for the distance between the two points given in last example, Art. (42).

17. Construct the point whose polar co-ordinates are, Art. (43),

$$u = 35^\circ, \qquad v = 70^\circ, \qquad r = 4.$$

18. Construct the right line, in space, whose equations are, Art. (44),

$$x = 2z + 3, \qquad y = -z + 2,$$

and find the equation of its third projection.

19. Find the point of intersection of the two lines, in space, whose equations are, Art. (47),

$$x = -2z + 3, \qquad y = z - 2;$$

$$x = 3z - 1, \qquad 5y = -10z + 2.$$

20. Find the expression for the cosine of the angle included between the lines given in the last example, Art. (49).

21. Ascertain if the lines whose equations are,

$$x = 2z + 1, \qquad y = 3z + 4;$$

$$x = -2z + 3, \qquad y = z - 2;$$

are parallel, or perpendicular, Art. (49).

22. Find the equations of a right line which shall pass through the point $x' = -3$, $y' = 2$, $z' = -1$, and be parallel to the line whose equations are, Art. (49),

$$x = -3z - 1, \qquad y = 4z + 3.$$

23. Find the equations of a right line which shall pass through the same point and be perpendicular to the same line, as in the last example, Art. (49).

24. Find the equations of a right line which shall pass through the two points, Art. (51),

$$x' = -1, \quad y' = 2, \quad z' = 0; \quad x'' = 3, \quad y'' = 0, \quad z'' = 2.$$

25. Find the intersection of the two lines whose equations are, Art. (53),

$$x^2 + z^2 - 5 = 0, \qquad z + y - 3 = 0;$$

$$x - 3z + 5 = 0, \qquad z^2 + 4y^2 - 8y = 0.$$

26. Find the equation of a plane whose directrix is represented by

$$4y - 3x + 1 = 0,$$

the projections of the generatrix making angles with the axis of Z, whose tangents are 2 and -3, Art (55).

27. Find the equations of the traces of the plane represented by

$$2z - 3y + x + 4 = 0,$$

and the points in which it cuts the co-ordinate axes, Art. (56).

28. Find the point in which the right line, whose equations are

$$x - 2z + 2 = 0, \qquad 2y + 3z - 1 = 0,$$

pierces the plane given in the last example, Art. (58).

29. Ascertain if the same line and plane are perpendicular to each other, Art. (59).

30. Find the equations of a right line passing through the point $x' = 1$, $y' = -3$, $z' = 0$, and perpendicular to the plane represented by

$$3x - 4y + z - 1 = 0;$$

also the point in which the line pierces the plane, and the length of the perpendicular, Art. (60).

31. Find the expression for the sine of the angle made by the line whose equations are

$$x = 3z + 5, \qquad y = -z + 1,$$

with the plane given in the last example, Art. (61).

32. Find the intersection of the two planes whose equations are, Art. (62),

$$3x - 5y + z = 0, \qquad x - y - 3z + 1 = 0.$$

Also, the expression for the cosine of the angle included by the same planes, Art. (63).

33. Find the equation of a plane passing through the origin of co-ordinates, and the two points, Art. (65),

$$x' = -1, \qquad y' = 2, \qquad z' = 3;$$

$$x'' = 0, \qquad y'' = -2, \qquad z'' = -1.$$

34. The equation of a circle being

$$x^2 + y^2 = 9;$$

find its equation referred to a system of co-ordinate axes, making an angle of 45° with each other, the new axis of X being parallel to the primitive, and the new origin being at the upper extremity of the vertical diameter, Art. (67).

35. Find the general equation of the circle referred to any set of oblique co-ordinate axes, Art. (67).

36. Find the general polar equation of the right line, Art. (69).

37. Find the equation of a cylinder, the equation of the directrix being, Art. (75),

$$y^2 = 2x - x^2,$$

and the elements being parallel to the line,

$$x = 2z + 4, \qquad y = -3z + 1.$$

38. Find the intersection of the cylinder of the preceding example by the plane whose equation is, Art. (62),

$$3x - 2y - 3z + 2 = 0.$$

39. Find the general equation of a cylinder, with an elliptical base, the origin of co-ordinates being at the centre of the base, Art. (75).

40. Find the equation of a cone, the co-ordinates of the vertex being $x' = 1$, $y' = 2$, $z' = -3$, and the equation of the directrix, Art. (77),

$$y^2 = 6x.$$

41. Find the intersection of the same cone by the plane whose equation is, Art. (62),

$$x + 2y - 3z = 0.$$

42. Find the equation of a right cone, the equation of the directrix being

$$x^2 + y^2 = 9,$$

the altitude being 5, Art. (77).

43. Intersect the same cone by a plane passing through the axis of Y and making an angle of 45° with the base, and find the equation of the intersection in its own plane, Art. (81).

44. Find the general equation of a cone with a hyperbolic base, the origin of co-ordinates being at the centre of the base, Art. (77).

45. Construct the parabolas whose equations are, Arts. (85), (86),

$$y^2 = 4x; \qquad y^2 = -3x; \qquad x^2 = 9y.$$

46. Ascertain whether the point $x' = -3$, $y' = 3$, is

without, on, or within each of the parabolas given in the preceding example, Art. (87).

47. Find the equation of a parabola which shall pass through the point $x' = 3$, $y' = 5$, Art. (29).

48. Find the intersection of the circle and parabola whose equations are, Art. (27),

$$x^2 + y^2 = 6, \qquad y^2 = 2x.$$

49. Find the equation of a tangent to the parabola $y^2 = -2x$, at the point $y' = 4$, $x' = 8$, Art. (90).

Find also the equation of a normal at the same point, Art. (98).

50. Find the equation of a tangent to the parabola $y^2 = 4x$, and parallel to the right line whose equation is, Arts. (30), (90),

$$2y = 3x + 5.$$

51. Find the equations of the two tangents to the parabola represented by $y^2 = 6x$, which shall pass through the point $x' = 1$, $y' = 4$, Art. (93).

52. Find the equation of the polar line to the point $c = 2$, $d = 1$, for the parabola represented by $y^2 = 3x$, Art. (95).

53. The equation of the polar line to the same parabola being

$$y = x + 2,$$

find the co-ordinates of the pole, Art. (95),

54. The equation of a parabola being $y^2 = 4x$, find its equation when referred to a diameter and tangent at its vertex, the tangent making an angle of 45° with the axis, Art. (99).

55. Determine the axes, and construct the ellipses, whose equations are, Art. (106),

$$2y^2 + 3x^2 = 4; \qquad 4y^2 + x^2 = 9.$$

56. Determine the axes and construct the hyperbolas, whose equations are, Art. (107),

$$y^2 - 3x^2 = -5; \qquad 2y^2 - 4x^2 = 4.$$

57. Ascertain whether the point $x' = 2$, $y' = 3$, is without, on, or within each of the curves given in the last two examples, Arts. (109), (110).

58. Find the equation of an ellipse which shall pass through the point $x' = 3$, $y' = 2$, the origin of co-ordinates being at the centre, and the semi-transverse axis equal to 4, Art. (125).

59. Find the equation of an hyperbola which shall pass through the point $x' = -3$, $y' = -2$, the origin being at the centre, and the semi-conjugate axis equal to 2, Art. (126).

60. Find the intersection of the ellipse and parabola, whose equations are, Art. (27),

$$2y^2 + 4x^2 = 8, \qquad y^2 = -5x.$$

61. Find the intersection of the ellipse and hyperbola, whose equations are, Art. (27),

$$3y^2 + x^2 = 3, \qquad 2y^2 - 3x^2 = -6.$$

62. Find the equations of a tangent and normal, to the ellipse represented by

$$4y^2 + x^2 = 9,$$

at the point $x'' = 1$; $y'' = \sqrt{2}$, Art. (128).

63. Find the equations of a tangent and normal to the hyperbola

$$4y^2 - 2x^2 = -8,$$

at the point $x'' = \sqrt{8}$, $y'' = \sqrt{2}$, Arts. (131), (30).

64. Find the equation of a tangent to the ellipse

$$4y^2 + 9x^2 = 36,$$

and making the angle 45° with the axis, Arts. (30), (128).

65. Find the equations of the two tangents to the ellipse represented by

$$4y^2 + 3x^2 = 12,$$

which shall pass through the point $x' = 1, \quad y' = 4,$ Art. (133).

66. Find the equations of the two tangents to the hyperbola represented by

$$y^2 - 3x^2 = -5,$$

which shall pass through the point $x' = 2, \quad y' = 3,$ Art. (134).

67. The equations of an ellipse and its polar line being

$$4y^2 + 2x^2 = 8; \qquad y = 2x + 6,$$

find the co-ordinates of the pole, Art. (139).

68. The equation of an hyperbola being

$$3y^2 - 2x^2 = -6,$$

find the equation of the polar line of the pole $c = 4, \quad d = 0,$ Art. (140).

69. Construct an ellipse, the two conjugate diameters of which are 6 and 4, making an angle of 120°; also an hyperbola having the same conjugate diameters, Art. (150).

70. Find the position and length of the equal conjugate diameters of the ellipse, whose equation is, Art. (159),

$$4y^2 + 3x^2 = 12.$$

71. Construct the asymptotes of the hyperbola,

$$4y^2 - 2x^2 = -8,$$

and find its equation when referred to them, Art. (161).

72. Construct the hyperbola whose equation is, Art. (170),

$$2xy + 3y + x - 1 = 0.$$

73. For examples illustrating the discussion of the general equation of the second degree between two variables, see Arts. (173), (176), (179).

74. Ascertain if the line represented by the equation

$$y^4 - x^2 - 2x - 3 = 0,$$

has a centre, and determine its co-ordinates, Art. (181).

75. For examples relating to loci, see Art. (194).

76. Find the equation of the surface generated by revolving the right line whose equations are

$$4x = 3z + 2, \qquad 2y = -z + 6,$$

about the axis of Z, Art. (196).

77. Find the equation of the paraboloid of revolution generated by the parabola represented by, Art. (198),

$$y^2 = -3x.$$

78. Find the equations of the spheroids, generated by the ellipse represented by

$$4y^2 + x^2 = 4.$$

79. Find the equations of the hyperboloids, generated by the hyperbola represented by

$$9y^2 - 4x^2 = -36.$$

80. Find the equation of the surface generated by revolving the parabola represented by

$$2y^2 = x + \text{[illegible]},$$

about the axis of Y.

81. Find the equations of the surfaces generated by revolving the lines represented by

$$y^2 = \frac{1}{x}, \qquad y^3 = 2x^2,$$

about the axis of Y.

Also the surface generated by revolving the first line about the axis of X.

82. Find the position of the planes which will make circular sections, Art. (226), in the ellipsoid whose equation is

$$2x^2 + 3y^2 + 4z^2 = 1.$$

83. Find the position of the planes which will make circular sections, Arts. (227), (228), in the hyperboloids whose equations are

$$x^2 + 2y^2 - z^2 = -3; \qquad 4x^2 - y^2 - 3z^2 = -2.$$

84. Find the position of the planes which will make circular sections, Art. (229), in the paraboloid whose equation is

$$2y^2 + 3z^2 - 4x = 0.$$

85. Find the equation of a tangent plane, Art. (232), to the ellipsoid, whose equation is

$$4x^2 + 2y^2 + z^2 = 10,$$

at the point whose co-ordinates are $x'' = 1$, $y'' = -1$, $z'' = 2$.

Also the equation of a normal line, Art. (236), at the same point.

86. Find the equations of the tangent planes, Art. (232), and normal lines Art. (236), to the hyperboloids whose equations are

$$2x^2 + y^2 - 3z^2 = -18; \qquad 3x^2 - 2y^2 - z^2 = -7;$$

at the point of the first, represented by $x'' = 2$, $y'' = -1$, $z'' = 3$; and at the point of the second, represented by $x'' = 2$, $y'' = -3$; $z'' = 1$.

87. Find the equations of the tangent planes, Art. (232), and normal lines, Art. (236), to the paraboloids whose equations are

$$2y^2 + 3z^2 = 4x; \qquad 4y^2 - z^2 = 5x;$$

at the point of the first, represented by $x'' = 5$, $y'' = 2$, $z'' = -2$; and at the point of the second, represented by $x' = 4$, $y'' = -3$, $z'' = -4$.

THE END

www.ingramcontent.com/pod-product-compliance
Lightning Source LLC
LaVergne TN
LVHW020237110826
845151LV00003B/950

* 9 7 8 1 4 2 5 5 2 9 8 1 9 *